Best Wishes.

GW01403421

ɔol.

~u/
Leicester
SCRCDM

Human Error – b

Simon Bennett

Perpetuity Press

Published by
Perpetuity Press Ltd

PO Box 376,
Leicester, LE2 1UP, UK
Telephone: +44 (0) 116 221 7778
Fax: +44 (0) 116 221 7171

214. N. Houston,
Comanche, Texas 76442, USA
Telephone: 915 356 7048
Fax: 915 356 3093

Email: info@perpetuitypress.co.uk
Website: http://www.perpetuitypress.com

First Published 2001

Copyright © 2001 Perpetuity Press Ltd

All rights reserved. Except for the quotation of short passages for the purposes of criticism and review, no part of this publication may be reprinted or reproduced or utilised in any form or by any electronic, mechanical or other means, now known or hereafter invented, including photocopying and recording, or in any information storage or retrieval system, without prior written permission from the publishersexcept in accordance with the provisions of the Copyright, Designs and Patents Act 1998.

Warning: the doing of an unauthorised act in relation to copyright work may result in both civil claim for damages and criminal prosecution.

The views expressed in this publication are those of the author and do not necessarily reflect those of Perpetuity Press Ltd.

British Library Cataloguing in Publication Data.
A catalogue record for this book is available from the British Library.

Human Error – by Design?
Simon Bennett
ISBN 1 899287 72 8

We can pull and haul and push and lift and drive,
We can print and plough and weave and heat and light,
We can run and race and swim and fly and dive,
We can see and hear and count and read and write...

But remember, please, the Law by which we live,
We are not built to comprehend a lie.
We can neither love nor pity nor forgive –
If you make a slip in handling us, you die!
- Rudyard Kipling, *The Secret of the Machines*

This research is dedicated to the vision and labours of all those,
from plane builders to pilots, who have created an industry of
global consequence – commercial aviation.

About the author

Simon Bennett is the Director of the Scarman Centre's distance-learning MSc in Risk, Crisis and Disaster Management. He has a PhD in sociology from Brunel University, London. His research interests include the socio-technical aspects of technological failure. He has published in numerous journals and periodicals including *Science, Technology and Human Values*, *The International Journal of Mass Emergencies and Disasters*, *The Log: Journal of the British Airline Pilots Association*, and the United Kingdom Flight Safety Committee's *Focus* magazine.

Contents

Preface

Air travel is a remarkably safe mode of transport. Consequently when accidents and disasters do occur they attract attention. The August 2000 crash of a Gulf Air Airbus into the sea off Bahrain was treated as 'breaking news' by the British Broadcasting Corporation. Accidents and disasters are frequently blamed on 'pilot error'. While there may be an understandable need for the media and public to ascribe blame, this trait may serve to hide the critical underlying cause(s) of disaster. Such causes may include inadequate aircrew training, badly designed controls or face-saving operational requirements (a noise-abatement flight profile, for example) that may, under certain unpropitious circumstances, reduce safety margins.

This monograph accepts that pilots can and do make mistakes. Nevertheless, it is argued that errors may also be 'induced' — by such factors as fatigue, bad design or incompetent air traffic control. In light of this the habitual blaming of pilots would seem unjust and, to the extent that 'blamism' obscures the underlying causes of error, dysfunctional in the matter of accident prevention. This monograph evaluates the utility of systemic or holistic analysis with regard to understanding the aetiology of aviation accidents and disasters.

Introduction

Disasters may be divided into two types: high probability, low consequence events, and low probability, high consequence events. A car crash typifies the former, an air crash the latter. Air crashes may be rare, but their outcomes can be severe. As Faith (1996a: 1) puts it; '... [T]he smallest detail, of design or maintenance, the smallest error by the pilot, can destroy the lives of hundreds of innocent passengers'. This monograph will examine the extent to which air disasters may be attributed *solely* to 'operator' or 'pilot error'. This research is undertaken in the context of contradictory views of the role of pilot error in aviation disasters. At one end of the spectrum of opinion are systems specialists like Shapero and Lee. In a survey of 9 United States Air Force (USAF) missile systems Shapero *et al*(cited in Greatorex and Buck, 1995: 176) found that between 20-53 percent of mishaps were attributable to 'human error'. Lee *et al*, reviewing studies of nuclear power, air traffic control and aviation discovered—according to Greatorex and Buck—that; ' ... 20-90% of system failures, depending on the application area, are attributable to the human element ... ' (1995: 176). At the other end are Shaw (1969) and Tye (1973) who assert that far fewer 'system failures' are attributable to the 'human element'—at least in the form of the *operator's* decisions and actions. The monograph aims to develop a fuller understanding of the relationship between human performance and such factors as the state of scientific and technological knowledge, industry regulation, training, relationships with work colleagues both during and after training, design, manufacture, testing, installation, maintenance and procedure, both in the matter of air safety and in the aetiology of air disasters. The larger object is to make a positive contribution to air safety at a time of rapid growth and development within the global aviation industry (Doganis, 1991: xiii; Tarry, 2000: 94).

Structure and method

The monograph commences with an exposition of what might be termed the 'conventional' view of the aetiology of air disasters—namely that most are *directly* attributable to pilot error. This is followed by a description of the counter-argument —that only a very small number of disasters should be blamed on pilot error. In the next section a theoretical perspective on human error is developed, using the work of academics like Professor of Psychology James Reason (1990) and Professor of Cognitive Science Donald Norman (1988; 1993). It could be said that Reason and Norman have developed an holistic or systemic view of disaster —one that postulates that the error that contributed to or caused the disaster may have been committed by an actor or by actors removed in time and space from the locus of the disaster itself. The following case studies, gleaned from secondary data, are used to illustrate the potentially complex and convoluted aetiology of aviation accidents and disasters. The views of Shapero and Lee, Shaw and Tye, Reason and Norman may — in some degree — be evaluated with reference to these studies. Finally some tentative conclusions are drawn regarding the role of pilot error in air disasters. (It is accepted that the limited number of case studies allows no more than this).

For the most part the monograph takes as its data base scheduled and inclusive tour (IT) commercial air operations. General and sport aviation are not considered in detail. Exceptionally, there may be reference to military aviation. For example one of the case studies concerns the ascription of blame following the crash during a training mission of a United States Air Force (USAF) aircraft.

The monograph focuses only on accidents and disasters: that is, events where there has been physical damage to the aircraft, possibly accompanied by injury and/or loss of life. It should be noted that while there are relatively few major accidents in aviation (Bordoni, 1997; Learmount, 2000a), there are many more 'incidents', where, as Barlay (1997: 35) puts it; '... [only] by the grace of some lucky element' did the episode not develop into a disaster. (Of course, that 'lucky element' may have been timely remedial action taken by the aircrew). Faith (1996b: 2) uses a familiar metaphor to describe the nature of the aviation safety issue. As he puts it; '... [C]rashes are the tiny tip of an iceberg composed of hundreds of thousands of less dramatic incidents involving an aircraft or its environment'.

Finally, and perhaps most crucially, the reader should be aware that the monograph has been written primarily as an accessible (that is, non-jargonistic and, hopefully, engaging) *introduction* to the application of systems thinking and holism to the investigation of aviation accidents and disasters. To this end it draws on established theory and, for the most part, well-known case studies dating back to the 1950s where, it can reasonably be assumed, most of the 'facts' have been revealed. If the monograph has a novelty, it lies in the fact that these several well-known cases have not been analysed in quite this way before. It is doubtful whether the 1958 Munich air disaster, for example, has, prior to the publication of this monograph, been systematically viewed through the theoretical frameworks of Sagan (1993), Buck (1994) and Horlick-Jones (1996). (The use of older case-study material provides an additional benefit for the author: it reduces the chance of litigation!).

Of course, not every writer in the field has been cited. Nevertheless an attempt has been made to include a representative sample of those who theorise and research aviation safety, from Hurst, published in the 1970s, to Snook, published in the new century. A conscious effort has also been made to cite (sometimes little-known) writers who render difficult concepts intellectually digestible (like Miller (1984)).

The author hopes that the monograph will be read by lay people and scholars, 'armchair' pilots and career flight crew, those who regulate and those who are subject to regulation. It has been written to appeal to all-comers, not just a select few.

Received wisdom

While there are national variations in levels of air safety (Beaty, 1991: 90; Faith, 1996b: 2; Barlay, 1997: vii-viii) and variations between carriers (Caesar cited in Barlay, 1997: 35) air travel, when considered 'in the round', is one of the safest means of transportation (Allen in Hurst, 1976: 93; Barlay, 1997: ix; Campbell and Bagshaw, 1999: 5-6). As Heino Caesar (cited in Faith, 1996b: 154), head of safety at Lufthansa (Germany's national airline) puts it; 'In Germany alone we have 8,000 dead on the streets every year ... worldwide we are killing 500 people in jet aircraft crashes'. Caesar's point is that, compared to the number of road deaths in just one country, commercial aviation is a remarkably safe mode of transportation. (In fact, the promiscuous traveller is 20 times more likely to be killed in a car accident than in an airplane accident (Buck, 1994: 5)). MacPherson, reviewing aviation safety in the United States, makes some interesting comparisons:

> In 1996 ... 394 people died in America [in commercial airline accidents] .... That 394 compares with 19,000 Americans that year who were murdered, 41,907 who succumbed in automobile accidents and 714 who died in boating mishaps (1998: 1-2).

According to Campbell and Bagshaw (1999: 6) the 1960s proved to be a 'watershed' decade for commercial air safety. As they explain:

> Prior to 1960, the annual accident rate for scheduled air carriers was something over 60 per million departures. In the ten years to 1970, this rate improved dramatically to about one or two per million departures, where it has remained.

Writing in the summer of 1999, Learmount composed the following optimistic assessment of commercial air safety:

> An airline industry-feared rise in air transport accidents is not happening .... Fatal accident data for the January-June sector of the past 10 years ... reflect an almost horizontal trend in both the absolute figures and the three-year moving average for a period in which passenger traffic has grown by nearly 50% .... The number of fatal accidents is absolutely level for the decade ... (1999: 31).

Looking back on the flight safety statistics for the 1990s, *Flight International* (2000a: 3) commented that '... the risk per flight and per passenger in the 1990s was slightly less than in the preceding decade ...'.

On the rare occasions when crashes do occur they are often ascribed to 'pilot error' or, more generically 'human error' (which can include errors made by second and/or third parties). A 1972 study by the International Air Transport

Association (IATA) found that '... forty-four out of sixty-three crashes [70%] which occurred during landing were attributed to pilot error' (White cited in Faith, 1996b: 153). A 1977 analysis of Federal Aviation Administration (FAA) reports '... showed that during a four-year period errors in pilot judgement accounted for over 50 per cent of pilot fatalities' (Beaty, 1991: 93). Masefield noted that; 'When we come to examine those [aviation] accidents that do happen ... we find that about half of them ... have been (and continue to be) ascribed to 'human error'' (Masefield cited in Hurst, 1976: v). In the 1995 *National Plan for Civil Aviation Human Factors* published jointly by the Federal Aviation Administration (FAA) the US Department of Defence (DoD) and the National Aeronautics and Space Administration (NASA) it was stated that; '... [T]he need for human factors research stems from the realisation that human error has been identified as a causal factor in 60 to 80% of air carrier, commuter and general aviation accidents and incidents' (1995: 10). A study of light aircraft accidents in Ontario, Canada, revealed 'pilot error' to be 'the most frequent cause of crashes'. Out of 47 impacts, 26 (55 per-cent) were attributed to pilot error (Shkrum, Hurlbut and Young, 1996). Studies by Boeing (the American aircraft manufacturer) and Lufthansa have pinpointed 'crew error' as the main factor in up to 75 per-cent of accidents (Barlay, 1997: 224). Perrow (1999: 133) has noted that in both commercial and military aviation between 50 and 70 per-cent of mishaps are ascribed to 'human error'. Writing at the end of the first century of powered flight, Weir (1999:140) noted that the aviation industry believed 'Pilot error [to be] the primary factor in over 70 per cent of all crashes .... In all accidents, human error, if not the primary factor, was involved in 85 per cent of crashes'.

Whichever figure one chooses to believe, it is clear that those in *immediate* control of an aircraft are often blamed for disaster. According to Allen (cited in Hurst, 1976: 93-98) this pattern may reflect the large areas of uncertainty that pervade commercial air operations. Aircraft are highly complex machines. Designers can never be certain that safety 'improvements' will lead to a net increase in safety levels. During *normal* air operations aircraft may be required to operate at the limits of their design capabilities in a volatile natural environment. According to Allen, such factors mitigate against the discovery of exactly what went wrong prior to a near-miss or accident. Under these circumstances it is unsurprising that the pilot — who, in theory at least, was 'in control' at the time — may be blamed for the disaster. As Allen puts it; 'It is this vast area of inevitable uncertainty and lack of pertinent data which leaves the residual doubt of the exact cause of an accident all too commonly ascribed to pilot error' (Allen cited in Hurst, 1976: 94).

There may, of course, be less 'honourable' reasons for the precipitous ascription of blame—intense media interest, for example (as with the Air France Concorde crash outside Paris on July 25, 2000), that might encompass the physical pursuit of key actors followed by demands for instant analysis, answers and solutions.

According to Allen (cited in Hurst: 1976: 93), disasters '... cannot but attract sensational press reaction'. Buck (1994: 4) shares this view, ascribing media interest to the *shocking* nature of disaster. As he puts it:

> An airline accident is a sudden, dramatic disaster, one that gets immediate media attention; before the smoke has cleared, the hounding questions heard from all quarters are: 'What happened? Who's at fault?'

Bressey (cited in Hurst, 1976: 60) talks about '... the insatiable need of the media for a readily identifiable scapegoat'. Under these circumstances, asserts Bressey, it may be difficult for those in official positions to resist the temptation to name those they *believe* to be responsible (Bressey cited in Hurst, 1976: 57-59). Referring to this phenomenon Bressey talks about '... the regrettable tendency on the part of the mass media — and even some authorities who should know better — to pass 'instant judgement' on the pilot, almost before the wreckage has stopped burning. In far too many cases, this off-the-cuff verdict of pilot error is returned on the flimsiest evidence' (cited in Hurst, 1976: 57). According to the veteran air-crash investigator Charles Miller (cited in Faith, 1996b: 3) post-disaster reporting and discussion are often one-dimensional, with both the media and public eschewing an holistic and multi-faceted analysis in favour of finding the 'smoking gun'. As he explains:

> I was so sick and tired of people taking that one thing and missing all the other lessons that I'm ready to tell 'em [sic], forget what caused the accident — report it all. Of course, the people in the media and the lawyers won't like that. *The idea of the single cause is a fixation of the media*, that's true, but I think you can almost go back to the lay public, who don't like to have complicated stories told to them (my emphasis).

Weir (1999: 140) makes a similar point in his assertion that '... many investigations stop analysing crashes once pilot error has been identified'. According to Barlay (1997: 120) the abbreviation of investigations may proscribe isomorphic learning (Toft's (1992) idea that safety can be improved by applying the lessons of disasters). As he puts it:

> Over-simplification precludes the discovery of the true causes of an accident and, above all, misses the opportunity to identify risks and so prevent disastrous repetitions.

Bressey (cited in Hurst, 1976: 60) opines that even if the temptation to pass instant judgement is resisted, the Public Inquiry may, as he puts it, 'devote inordinate effort' to the apportionment of blame in preference to establishing the circumstances and socio-technic aetiology of the disaster. According to Weir (1999: 149); 'The attaching of blame to pilots can be the result of pressures brought about by the legal system'. Weir suggests that '... strict legal responsibility and fault-finding can obscure the reasons for a crash'. He describes the benefits of the alternative, open, non-blamist approach as follows:

... [O]pinion tends to diverge into two schools of thought: one says blame or negligence is an important concept legally and helps prevent people ducking their responsibilities. The other school says that assigning blame gets in the way of identifying the true sequence of accident causation and thus preventing repetitions (1999: 149).

According to Santilli (cited in Perrow, 1999: 133) the attaching of blame to pilots may reflect a desire to cover up incompetence elsewhere in the aviation system or failure to pinpoint the exact cause(s) of disaster.

In his book *Friendly Fire* Scott Snook (2000) alleges that the blaming of the US fighter pilots who shot down two allied helicopters over Northern Iraq concealed a litany of more fundamental organisational weaknesses, from role confusion within the Airborne Warning and Control (AWACS) aircraft circling overhead at the time to poor co-ordination within the US military. Snook uses the term 'fundamental attribution error' to describe the process whereby pilots are blamed regardless of context. As he puts it; 'The fundamental attribution error lives. In spite of repeated warnings to the presence of this human frailty, when it comes to accidents such as this shootdown, our desire to blame the 'man in the loop' remains overwhelming.... No matter how often we are reminded of our natural inclination to blame the individual (read: 'others'), no matter how often these attributions turn out to be false, the simple fact remains: our first impulse is to point the finger at the nearest person when something goes wrong. When it comes to explaining the cause of accidents, pilot or operator error remains the single most common attribution—by far' (2000: 205).

Perhaps understandably, the precipitous ascription of blame—often to pilots— causes resentment within the flying community. As Masefield (cited in Hurst, 1976: vii) explains:

> ... [T]here exists a clear undercurrent of natural resentment among active pilots ... that the blame for pilot error ... is always lurking just around the corner *and has been so often expressed* (my emphasis).

Holmes (cited in Barlay, 1997: 114) comments that the recording of flight deck conversations by cockpit voice recorders and the subsequent dissemination of transcripts can lead to (sometimes badly injured) pilots being '... found guilty by editorials, with no effective right of reply'.

It must be said, however, that, following Buck's (1994: 5) argument, there may be a certain irony in expressions of resentment by the industry at the conduct of the press. As Buck puts it:

> One might theorise on the ironic possibility that aviation's courtship of the media for banner headlines to reward its great feats plays a part in the lurid response to accidents by newspapers and TV.

One might conclude that in aviation, as in so many other fields of human endeavour, those who live by the press may, under certain conditions, suffer at the hands of the press.

It must also be recognised that even the most highly-trained and accomplished pilot can err (Campbell and Bagshaw, 1999: 6; Perrow, 1999: 133; Snook, 2000: 205). In 1994 an Aeroflot Airbus crashed killing 75. The Captain had allowed his fifteen year old son to fly the Airbus. His son had lost control, and the crew were unable to recover the aircraft (Brookes, 1996: 129-130). Pilots have also been known to use their aircraft to commit suicide (Cullen, 1998). Less spectacularly, a study of pilots flying short-haul commercial routes found that crews committed at least one error every four minutes. According to Perrow, however; 'The vast majority of these errors [were] caught very quickly, or [were] insignificant' (1999: 133).

Without wishing to deny the problem of pilot error *per se*, it should be remembered that the industry itself may exhibit poor quality control of aircrew skills and standards. As Learmount (2000b: 9) explains, a recently published Safety Assessment of Foreign Aircraft (SAFA) report listed such omissions and errors as 'non-valid flight-crew licenses, the absence of required manuals, or manuals being out of date, and incorrect calculation of load distribution'. Having said this, however, many hold the view that most pilots are competent and rigorous in their approach. According to Bressey (cited in Hurst, 1976: 13) pilots are '... rigorously selected, highly trained, disciplined and vocationally dedicated'. Masefield (cited in Hurst, 1976: vi) describes them as a 'dedicated band of hard-working technocrats'. Weir (1999: 141) opines that pilots '... take pride in their ability to operate the aircraft as efficiently and comfortably as possible'. Weir describes their 'personal commitment' and makes reference to 'a high level of continual and refresher training and medical examinations' within the industry.

While Weir's portrait of a conscientious industry striving to maintain standards might be challenged by some, there is evidence that the industry, perhaps with a little prompting, is doing the best it can in an economic climate that ranges from slump to helter-skelter expansion (Doganis, 1991: xiii-xiv). The Asian and Pacific Rim airlines are a case in point. In recent years a number of Asian-Pacific airlines have suffered what might be termed an unacceptable level of accidents and disasters (Learmount, 2000a: 2-3). Part of the problem lay in the rapid expansion of the 'Asian Tiger' economies. As demand for commercial air transport services grew the region's airlines found themselves over-stretched and unable to resource their operations adequately. Training standards slipped. After a number of high-profile accidents '... the government stepped in and reined back growth so that training and infrastructure development could catch up. Safety improved markedly' (*Flight International*, 2000: 3).

Perhaps the final word on the quality of aircrew should be left to two experienced commentators, Captain Heino Caesar of Lufthansa and Robert Besco, a professional pilot. Caesar (cited in Faith, 1996b: 154) ventures the opinion that 'In hundreds of thousands of cases pilots have prevented an accident by their action'. Besco (cited in Faith, 1996b: 154) offers the following assessment:

> I hardly think of them [pilots] as a breed, although I can see some common characteristics. First, they are probably brighter on average than people in most professions. It takes a lot of cognitive ability to be a pilot. The capacity to stay focused on your primary objective when things are going bad around you, and to be able to operate under stress, are probably the two main characteristics of pilots.

In the context of this discussion about 'received wisdom' something also needs to be said about the hardware manufacturers—those corporations that actually make the aircraft, engines and ancillary equipment that, when married together, constitute a Boeing 747-400 or Airbus Industrie A340. No-one should be in any doubt about the scale of achievement of the world's airplane manufacturers. In the 1960s Boeing gambled its very existence on the development of an entirely new type of aircraft, the large wide-body. The gamble paid off, and the 747 has become one of the world's best-selling and most popular aircraft. In the same decade the British and French designed and built a supersonic transport aircraft (SST) that operated for many years without a hull-loss. Boeing were able to bring their aircraft to market in a relatively short space of time. This was due in no small part to the enthusiasm, flexibility and technical competence of their workforce (DD: 1991). The fact that Concorde could be designed and built via a lengthy collaboration between two national aircraft industries with different languages and systems of measurement is further testimony to the expertise and dedication of those who work in the aircraft manufacturing sector. Given such achievements it might reasonably be assumed that airplane manufacturers make every effort to apply the lessons of component failure. While this may well be the 'received wisdom' within and without the industry there is some evidence to suggest that manufacturers do not always capitalise fully on experience. On March 3, 1974, a McDonnell Douglas DC-10 (a wide-bodied passenger jet) crashed near Paris. Three-hundred and forty-six died. An improperly-secured (and, some would say, poorly designed) cargo door had blown open at approximately 12,000 feet (the act of pressurising an aircraft stresses its skin), causing the passenger cabin floor to collapse, which severed the aircraft's control cables and hydraulic lines (the first generation of wide-bodied aircraft, of which the DC-10 was an example, had mechanical control linkages (which, in the DC-10, ran under the cabin floor). Today's 'fly-by-wire' aircraft have electrical linkages). Out of control, the DC-10 hit the ground at speed. According to Perrow (1999: 139-140) this failure mode had been presaged on a number of occasions;

> McDonnell Douglas had been warned of this [failure mode] by a Dutch engineer in 1969 ... by a McDonnell Douglas subcontractor ... in the spring of 1970 ... by a static ground test of the airplane in May ... 1970 ... by eleven entries in maintenance

logs up to June 11, 1972, concerning difficulties with the door; and by a narrowly averted disaster near Chicago on June 12, 1972, when the cargo door blew out.

It would be incorrect to assume that McDonnell Douglas did nothing, however. The company modified the design after the static ground tests (although, according to Perrow, this modification was 'inadequate'). Following the Chicago incident, North America's Federal Aviation Administration, at the request of the National Transportation Safety Board, agreed a 'nonmandatory' modification to the cargo door locking mechanism. Unfortunately, the Turkish Airlines DC-10 that crashed near Paris had not received this modification. According to Toft and Reynolds (1997: 62); 'It should ... be noted that prior to the Turkish Airlines disaster different operators had filed to McDonnell Douglas over a hundred reports of problems regarding the closing of cargo doors on the DC-10, many of them describing the same kinds of faults'. Of course, one might reasonably assume that over the past thirty years procedures and levels of accountability within the manufacturing sector have been improved. On January 20, 2000, both cowling shells on the port (number 1) engine of an Airbus A320 passenger aircraft flew open on take-off. Despite structural damage the airplane landed safely. This was not a novel problem on A320s fitted with International Aero Engines V2500 engines. According to *Flight International* (2000b: 15) the subsequent UK Air Accident Investigation Branch report criticised Airbus '... for failing to correct a problem which led to seven cowling separation incidents on A320 series aircraft'. It might be argued, on the basis of such evidence, that manufacturers do not always investigate component or other failures in sufficient depth, or, if they do, that they do not always act in time — or act at all. While it is accepted that investigations take time (no-one would wish to implement a fix that did not work or that eroded safety further) the fact remains that accidents repeat.

In conclusion, whatever the aetiology of an air crash and whatever the calibre of the aircrew, it is a fact that those in direct and immediate control of the aircraft are often picked out as most blameworthy. As Buck (1994: 3) explains; '... [W]hen an accident is labelled human error—and 60 percent to 90 percent are, depending on who you're listening to—the thoughts and focus go right to the cockpit, and pilots still collect the major blame'. According to Faith (1996b: 153) judgement can be passed quickly, and on little or no conclusive evidence:

> ... [P]ilot error ... comes automatically to mind .... Before any other explanations have been sought, pilots can be blamed for their incompetence, their casualness, or their characters.

The challenge

Despite the existence of a 'blamist' culture in society at large, in recent years certain parties have come to accept that factors other that simple 'operator error' may impact a course of events. As Allnutt (cited in Hurst, 1976: 65-66) explains:

... a statistical evaluation (Wansbeek, 1969) of 83 pilot-error accidents produced the following table of causes: Improper flight technique 15 ... Over-confidence 16 ... Insufficient care 18 ... Other causes 34 .... Such categorisations have long been established as research ground-lines. Pilot error has been defined, for example, in terms of failures of co-ordination and technique; or, as a consequence of shortcomings in the exercise of alertness and observation, or intelligence and judgement; or again, as a phenomenon influenced by personality or temperament (McFarland, 1953). Similarly, *other analyses list among accident causes those which may have been induced by design factors or operational procedures,* by ignorance, by deliberate acts of omission or commission, *by environmental factors,* or by psychological or physiological causes (my emphases).

Faith (1996a: 1) supports this more inclusive and systemic/holistic analysis of the causes of disaster. As he explains:

> Nature, in the form of freak, or simply bad, weather has always been a major factor, but *most crashes are traceable to some flaw in the human chain,* either among those most directly involved — aircraft designers and manufacturers, maintenance engineers, pilots and crew, air-traffic controllers and airlines — or among more shadowy figures, such as the authorities who control the development of new aircraft, flight patterns and the discipline imposed on pilots (my emphasis).

Faith's assertion that crashes often lie at the end of a 'chain' of actions is echoed by Weir (1999: 148) in his assertion that '... pilot error is usually only one part of a confluence of events in an accident ...' and by Snook (2000: 205) who ventures that 'an individual action' may be considered 'the final link in a long chain of events'. The notion that accidents are events with complex, possibly non-linear aetiologies resonates with Wagenaar and Groenewold (1987). Horlick-Jones (cited in Hood and Jones, 1996: 62-63) describes the argument as follows:

> [T]he causal link between the trigger event and the 'system failure' may be extremely complex. Wagenaar and Groenewold ... describe such failures as 'the consequences of highly complex coincidences' .... [H]uman error ... can trigger ... disasters that ultimately occur for very complex reasons. In a sense the impact of a simple error can be 'amplified' by ... sociotechnical context .... [T]he individual who makes the mistake [may be] caught in what one might call a 'systemic net' of circumstances beyond their control.

The notion that accidents may have a multi-factoral aetiology is supported by Barlay (1997: 119-120). Barlay challenges what he calls the 'over-simplification' of enquiry. As he explains:

> If in doubt, blame the pilot — that used to be the simple answer to puzzling mysteries of the air. Baffled aircrash investigators resorted to it, licensing authorities, manufacturers and others potentially involved in disasters hid behind it .... That simplistic early answer was unhelpful. It only disguised the circumstances that had *ensnared* the pilot. For a better understanding of errors, the term 'human factor' was coined. It saw the *pilot as a part of a system* that

allowed, facilitated, even invited errors, and encompassed error prone human elements in aircraft design, air traffic control, airport management, airline procedures, training, etc (my emphases).

Barlay's view that a system may be so designed as to invite error resonates with Weir (1999: 140) who asserts that '... manufacturers often build machines which are inviting operator errors'. Buck (1994: 3-4) produces the following summation of the numerous factors that, in his view, may — in varying degrees and through various interactions at different junctures — contribute to or cause a disaster:

> Human error is seldom accounted for in its entirety. This flying business is a big, complicated arena with many humans involved, most of whom are seldom if ever mentioned. To point out a few: the people who design and manufacture the airplane who may have made a serious error in stability and control; the test pilots who may not have been firm enough in condemning the fault; the company disregarding the pilot's warning; the Federal Aviation Administration (FAA), who makes the rules and puts its stamp of approval on the aircraft by granting its license; the air traffic control (ATC) system, whose employees are made to work with the concept that they must attempt to shove two pounds in a one-pound bag on a daily basis; the US Congress for legislation passed in the heat of emotions or for funding unprovided; even the President of the United States, who upset the ATC system by firing experienced air traffic controllers who were impossible to replace for years thereby screwing up an already overburdened system [after a strike in 1981 President Reagan fired 11,400 controllers]. There are also passengers, who add pressure unknowingly with their gripes and demands to make schedules for their connections, as well as their unspoken demands for a smooth ride, which pressure the pilot to request and at times argue with ATC for a better altitude where the air is smooth; the flight instruction system that taught the pilot to fly but missed on teaching judgement; the mechanic who didn't secure an oil drain plug; the people behind the weather dissemination system who didn't alert the pilot to a severe and growing thunderstorm at the airport of destination; the people responsible for airport design and construction who create a mishmash of taxi strips difficult to follow or runways located so aircraft must cross them while taxiing, which in turn strains the babble of communication to the point where misinterpreted instructions can put an airplane on an active runway in position for collision; the members of the local planning commissions who allow schools to be built almost on the end of a runway and then demand that the pilot follow hazardous noise reduction procedures; and the executives of the airlines who pressure the ATC system with more flights than the system can handle.

Like Buck, Barlay, too, highlights the potential impact of poor regulation and deficient airframe maintenance on levels of air safety:

> Inadequate training, resources, regulations and supervision cause numerous mishaps, and while the human factor in the cockpit continues to bear much of the responsibility, it is only now that western experts have begun to examine seriously the human factor in maintenance, too (1997: viii).

According to Allen (cited in Hurst, 1976: 92) Barnaby's book *Some Ship Disasters and their Causes* was one of the first studies to draw attention to the propensity of official inquiries to habitually allocate blame to the person in charge at the time of the disaster — whatever the aetiology of the event. After investigating almost one hundred cases of collision and grounding, Barnaby concluded that, as Allen puts it; '... the ship's captain was accorded a manifestly unjust proportion of the blame'. Allen goes on;

> ... there is a consistency in the philosophy which shaped the fiercely autocratic judgements of the Admirals who conducted those Courts of Enquiry. In their opinion ... command — i.e. the status of Captain — is equated with absolute responsibility, *regardless of inadequate design or provision* (cited in Hurst, 1976: 92).

Barnaby's analysis, published in 1968, challenged the received wisdom that the person or persons in charge at the time of a disaster must carry the burden of responsibility for that disaster — regardless of the historical background, the social, economic and/or political circumstances that obtained at the time or the technical aetiology of the event. In 1973 the Civil Aviation Authority's (CAA's) Controller of Safety, Dr Walter Tye, used the same argument as that deployed by Barnaby in a lecture to the Royal Aeronautical Society (RAeS). During the lecture Tye questioned the utility and morality of the prevalent 'blamist' culture:

> Human failures are not usually blameworthy, as they stem from insufficiency of knowledge or skill or foresight .... I would prefer not to label these accidents as crew error. Rather I would view them as failure of design, or lack of provision of aids, or inadequate training or planning, which *allowed* human error to occur too readily .... I would not write off crew error accidents as unavoidable, but view them as accidents which, with better design of aircraft, instruments, ground aids, training or procedures are amenable to reduction (Tye cited in Hurst, 1976: 63-64, my emphasis).

Tye's view that good design is a bulwark against mishap is supported by, for example, Allen. As he explains:

> The major stages in the creation of an aircraft are related together in a framework within which a great variety of choices must be taken, and an equally significant level of compromise arrived at. It is in establishing these decisions ... *that the seeds may be sown for future errors and accidents* (Allen, cited in Hurst, 1976: 96).

Masefield (cited in Hurst, 1976: v) opines that while '... every accident is a result of human error', the error that causes or contributes to an accident may be made by an actor — perhaps a designer of mechanical components or computer software — removed in both space and time from the event:

> The truth is that in some degree or other every accident is a result of human error, whether that error be direct — as, for instance, a pilot error on landing — or indirect, as in, for instance, an insufficiently thought-out design feature which has led to a

catastrophic failure of some component of an aeroplane or of its systems. Every mechanical process is potentially vulnerable to some human mistake or misjudgement, however far back along the line that failure may have been.

Of course, while it may be axiomatic that good design can act to improve safety, the realisation of this ideal may prove difficult, if not impossible, due to the cost of new developments in an industry that may be subject to powerful economic and competitive forces (Allen cited in Hurst, 1976: 94). According to Barlay (1997: 38) such forces may act to erode safety margins. As he explains:

> Deregulation, cut-throat competition and the price wars tend to force even the most safety-conscious manager to make compromises. If he [sic] *actually* compromised safety, he would be fired. But it is hard to detect when he goes just to the brink with hardly visible corner-cutting under the banner of *economising* that may in the end erode the safety cushion protecting the aviation industry.

Ironically one of the strengths of aviation design and technology may, under conditions of severe competition, prove to be a weakness. Thus, as Wittorf (cited in Barlay, 1997: 38) explains, the duplication of mechanical and avionics systems means that maintenance staff, in order to meet a repair deadline or budget target, may allow an aircraft to fly with error-prone systems in the knowledge that — provided the flight crew do not identify the problem during pre- and post-flight checks — such duplication will 'hide' the problem until such time as schedules and/or budgets allow remedial action to be taken. Of course, such expediency acts to reduce the overall safety margin both for the aircraft itself and for the air traffic system as a whole.

Safety margins are 'constructed' through a complex series of decisions and actions that impact systems reliability. Systems reliability is a function of timely and insightful design, competent manufacture from high-quality materials, adequate testing, correct installation and integration with other systems (both mechanical and electrical), comprehensive training of both cockpit and cabin crew and maintenance staff, and ongoing, high-quality planned and ad-hoc maintenance. All these safety-enhancing activities cost money. Where avionics and other systems are concerned, quadruplication is safer than triplication. Again, however, there are cost implications. As Allen (cited in Hurst, 1976: 114) explains; 'Clearly the provision of duplicated systems and time-consuming ground tests increases costs .... Improving the reliability of equipment leads to additional first cost, and to further testing cost .... [R]ather than reducing with time, both flight-test duration and cost show a steady increase ... in 1963 the Boeing 727 [a medium-haul passenger jet] ... needed a flight testing time of 1,137 hours .... By 1969 the ... Boeing 747 [a long-haul passenger jet] ... logged a total test time of 1,444 hours'. Concorde's designers anticipated a 4,000-hour test programme with the project's prototype and pre-production airframes (which still failed to prevent the crash of an Air France Concorde in Paris and a number of chronic defects (Kingsley-Jones, 2000: 54)).

Of course the quality of a flight-test programme reflects both the number of test hours flown and the relevance of the tests themselves. To the extent that flight tests are based on an accurate model of the performance of the aircraft's myriad components, they are valid. Should, however, those tests be informed by erroneous assumptions, they are not only invalid, but potentially dangerous.

In conclusion, in recent years there has been a realisation that poor design (of both hardware, software and standard operating procedures) and/or inadequate manufacture, testing, installation, maintenance and training (of both air and ground-crew) may either contribute to or cause 'pilot error'. It has been claimed that better design and improvements in the reliability of equipment can have a positive impact on the safety of air operations. While few would claim that pilots are blameless in the matter of near-misses and accidents, it is important, both in terms of ensuring flight safety and the just apportionment of blame, that the *underlying* causes of error are understood before judgement is passed. Perhaps the most dramatic challenge to the received wisdom that the majority of mishaps are caused by pilot error was made in 1969 by Dr R.R. Shaw of IATA. In an address to the Society of Licensed Aircraft Engineers and Technologists, Shaw (cited in Hurst, 1976: 137) claimed that the 'wilful neglect of procedure' by pilots accounted for 'one or two per cent of accidents—not sixty per cent'. As he put it:

> Apart from the rare exceptions, pilots are a ... responsible group of men [sic], and pilot error accidents occur, not because they have been sloppy, careless or wilfully disobedient, but *because we on the ground have laid booby traps for them, into which they have fallen.*

There may be some truth in Shaw's analysis: in 1973 a passenger aircraft crashed near Basel. One hundred and eight passengers died. The Captain and First Officer were found to be 'in error'. They had misinterpreted radio signals. However, as the crash report (cited in Hurst, 1976: 137) illustrates, there were mitigating circumstances:

> ... The crash was caused by faulty reception of navigational radio signals. This resulted in a pilot's error .... The imperfect reception was due to atmospheric disturbances, but it may be been [sic] compounded by defective equipment aboard the plane.

Thus while the *immediate* cause of the accident was the crew's misinterpretation of a signal, the *underlying* causes were a volatile natural environment and, possibly defective, equipment—factors over which, it could be argued, the Captain and First Officer had little or no control. The volatility of the physical environment in which aircraft operate may have a major impact on the safety of air operations. As Faith (1996b: 2) explains; '... [The] environment is hostile. In the words of the Flight Safety Foundation, 'Aviation in itself is not inherently dangerous, but, like the sea, is terribly unforgiving of any carelessness, incapacity or neglect''.

In the context of the received wisdom within the aviation industry that the majority of accidents (and, in one view, over 70% of accidents) are attributable to pilot error, Shaw's assertion that as few as two per cent of accidents are attributable to the 'wilful neglect of procedure' by pilots constitutes a controversial hypothesis. One might say that Shaw stands at one end of the 'spectrum of causation', with theorists like Shapero and Lee, agencies like the FAA and aero-industry corporations like Lufthansa and Boeing inhabiting the opposite polarity. This monograph will use a case-study approach (albeit a very small number of case studies due to the limitations of the format) to test the view that such factors as poor design, inadequate training, imperfect knowledge, overly restrictive and therefore potentially dangerous standard operating procedures (SOPs) and unrealistic economic expectations may play a larger role in the aetiology of disaster than might first appear. As Vaughan (1996: 463) argues in her seminal work *The Challenger Launch Decision*; 'What matters most is that we go beyond the obvious and grapple with the complexity, for explanation lies in the details'.

A systemic approach to understanding 'human error'

While neither Reason nor Norman would disagree with the assertion that a high proportion of accidents are caused by human error, both would take issue with the view that most accidents are caused by operator error—that is, by the person notionally 'in charge' at the time of the event. Reason's view that many accidents are caused or catalysed by persons not present at the time of the event is premised on the assumption that there are two quite different types of error—'active' errors and 'latent' errors. The former are committed by operators. The latter are committed by persons removed in time and space from the locus, both temporal and physical, of the event in question. In Reason's opinion it is the latter type of error that poses the greatest danger to such complex systems as modern commercial or military aircraft. As he explains:

> ... [L]atent errors pose the greatest threat to the safety of a complex system. In the past ... accident investigations have focused primarily upon active operator errors and equipment failures. While operators can, and frequently do, make errors in their attempts to recover from an out-of-tolerance system state, many of the root causes of the emergency were usually present within the system long before these active errors were committed. Rather than being the main instigators of an accident, *operators tend to be the inheritors of system defects created by poor design, incorrect installation, faulty maintenance and bad management decisions. Their part is usually that of adding the final garnish to a lethal brew whose ingredients have already been long in the cooking* (1990: 173, my emphasis).

In Reason's view, such factors as poor design, sloppy installation, inadequate maintenance or ineffectual training, either singly or in combination, may offer what he terms an 'affordance' for error. In effect, such factors may either cause, catalyse, facilitate or induce operator error. Norman (1988) supports this view of disasters as having potentially a multi-factoral aetiology. Thus, in his book *The Psychology of Everyday Things*, Norman mused that many everyday objects had been designed with little thought for the capacities of the user. As he puts it:

> The truth emerged slowly. My research activities led me to the study of human error and industrial accidents. I began to realise that human error resulted from bad design. Humans did not always behave so clumsily. But they do so when the things they must do are badly conceived [and] badly designed. Does a commercial airliner crash? Pilot error, say the reports.... But careful analysis ... usually gives the lie to such a story (Norman cited in Reason, 1990: 235).

According to Reason (1990: 197) systems may contain a number of badly designed elements (these can range from mechanical components to decision-support software to training and maintenance programmes to SOPs) which, either singly or in combination may, under 'propitious' or 'fertile' circumstances, facilitate or induce operator error. Reason calls these elements 'resident pathogens'. As he explains:

> The resident pathogen metaphor emphasises the significance of causal factors present in the system *before* an accident sequence actually begins. All man-made systems contain potentially destructive agencies, like the pathogens within the human body. At any one time, each complex system will have within it a certain number of latent failures, whose effects are not immediately apparent but that can serve both to promote unsafe acts and to weaken its defence mechanisms. For the most part, they are tolerated, detected and corrected, or kept in check by protective measures (the auto-immune system) [or, in the case of an aircraft, duplicated or triplicated systems]. But every now and again, a set of external circumstances — called here local triggers — arises that combines with these resident pathogens in subtle and often unlikely ways to thwart the system's defences and to bring about its catastrophic breakdown (1990: 197, my emphasis).

Reason asserts that 'major disasters' are rarely attributable to a single factor. Rather they are caused by a combination of factors 'either mechanical or human' that, through serendipity, interact in such a way as to produce instability. As he puts it; '... [T]hey arise from the unforeseen and usually unforeseeable concatenation of several diverse events, each one necessary but singly insufficient' (1990: 197). In other words, Reason believes most disasters to have a multi-factoral aetiology. It follows that the frequent ascription of culpability to a very small number of people — often the operators — is both illogical and unjust. It is also short-sighted, as it means that the true causes of and reasons for a near-miss, accident or disaster may never be understood, which obstructs isomorphic learning, which obstructs efforts to build safer socio-technical systems. Fitzgerald (1968) describes both the moral and efficiency dimensions to the issue of culpability. Hood and Jones (1996: 62) paraphrase his argument thus:

> Fitzgerald ... has noted that there is a general principle which states that a person should not be punished for occurrences over which they could exercise no control. He observed that to penalise such an individual [an operator who errs] is not only unfair but also inefficient, 'because it would not prevent similar occurrences in the future'.

It is clear from Reason's discourse that operator error may be induced by bad design. In his book *Things That Make Us Smart* Norman (1993) suggests that bad design may be rooted in a slavish devotion to science and technology and over-indulgence of the Modernist technocratic agenda. According to Norman this tendency found its most assertive expression in the slogan coined by the organisers of the 1933 Chicago World's Fair; 'Science Finds, Industry Applies, Man Conforms'. Such 'techno-centrism' is anathema to Norman, who makes a determined case for the 'human-centred' approach to design. As he explained in a 1997 critique of *Things That Make Us Smart*; 'The theme of the book is that technology can indeed enhance human intelligence, but only if it is properly built to fit human abilities and needs. Alas, all too often it isn't. All too often it is people who must conform to the technology. The proper way is, of course, for technology to conform to people .... I try to be a fan of technology: too bad the technologists get in the way' (Norman, 1997). As a riposte to the Chicago World's Fair slogan, Norman offers a new mantra for his proposed 'human-centred' era of design; 'People Propose, Science Studies, Technology Conforms' (Norman cited in Kirkus Reviews: 1993). The theme of human-centred design has been picked up by Greatorex and Buck who, like Norman, ask that scientists, technologists and designers give *primacy* to human capacities when developing novel control systems and/or artefacts:

> [T]he human-centred design approach requires information about human capabilities and task requirements to be in place prior to design. Too many systems have suffered premature design commitment and a great deal of money has had to be spent massaging technology-centred designs into a usable form .... It is time to implement user-centred design procedures so that the demands of systems do not exceed user capabilities .... [I]t is important to determine the demands on the user for system tasks and to match them to user capabilities .... *[T]he mental ... load placed upon the user by the system should be controlled and managed just as hardware power and weight budgets are maintained* (1995: 181-185, my emphasis).

The potential contribution of the 'human-centred design approach' to safety has also been recognised by the aviation press, as the following Editorial from *Flight International* demonstrates:

> If ... manufacturers do not want to be continually sent back to the drawing board they are going to have to put the pilot at the centre of their thinking, despite the increasing automation of cockpits .... *[P]ilots must be given cockpits in which they will make fewer equipment-led mistakes* ... (1999: 3, my emphasis).

According to *Flight International*, the human-centred approach to design had its origins in the Vietnam War, when the US military found that contractors, working to official requirements and specifications, were developing weaponry '... so complex that it was beyond the ability of soldiers to use it without massively expanded training'. According to *Flight International*; 'Basically, the human part of the system was not being given enough consideration' (1999: 3). One of the products of the dissonance between designer and user was a novel discipline — that of 'human-centred engineering', in which attention was focused on the 'man-machine [sic] interface'. Human-centred engineering examines '... ease of use, the effects of error during use, task distribution, and the adequacy of feedback to the user ...' (*Flight International*, 1999: 3). The objective is partly to keep the user cognitively engaged with the technology and its systems so that, should the need arise, s/he can intervene to correct whatever out-of-tolerance system state has developed (a stall situation, for example).

This, of course, is the ideal. However, as Reason and Norman remind us, we have not yet attained this 'ideal state' of systematic, global, inter and intra-disciplinary human-centred design. Operator errors can and do originate in poor design decisions. They also originate in necessary post-design activities; that is, in operations like manufacture, installation, training or maintenance. These *remote* decisions and/or activities, over which the operator has no direct or, perhaps, even indirect control, can generate an affordance for error. This affordance may or may not be realised by the operator. Reason defines the concept of affordance thus:

> The term affordance refers to the basic properties of objects that *shape* the way in which people react to them (1990: 235, my emphasis).

In Reason's opinion, therefore, the physical and/or logical properties of an object or system may *induce* a particular behaviour or reaction. When this behaviour or reaction is inappropriate to the circumstances obtaining at the time, a near-miss, accident or disaster may result.

In looking beyond the temporal and physical locus of an event, both Reason and Norman could be said to be taking an holistic or *systemic* view of the aetiology of the near-miss, accident or disaster. According to Blockley (cited in Hood and Jones, 1996: 32) the concept of systemicism is being deployed 'more and more often'. There may be some truth in this assertion given *Flight International's* optimistic prognosis for user-centred design (reproduced above). Blockley defines what he terms 'systems thinking' as follows:

> 'Systems' thinking ... is more an approach than a topic, a way of going about a problem. In simple terms, the systems approach is one that takes a broad view, which tries to take all aspects into account and which concentrates on the interactions between different parts of the problem. Some of the key concepts in this approach are world-view [and] holons ... (cited in Hood and Jones, 1996: 32).

Given this definition it might be argued that the Public Inquiry held in the wake of the 1987 *Herald of Free Enterprise* capsize outside the Belgian port of Zeebrugge adopted a systemic or holistic approach to its investigations. Thus while the Inquiry, chaired by Mr Justice Sheen, found the *immediate* cause of the disaster to be the Assistant Bosun's failure to close the vessel's bow doors as it exited the harbour — the crucial 'operator error' — it also included within its purview what it considered to be the underlying causes of disaster. As the Inquiry report put it:

> The underlying or cardinal faults lay higher up in the company. The Board of Directorships [sic] did not appreciate their responsibility for the safe management of their ships (and) did not have any proper comprehension of what their duties are .... From top to bottom the body corporate was infected with the disease of sloppiness (cited in Hood and Jones, 1996: 64).

Thus the Inquiry Report suggested that 'front-line' operators, far from being 'free agents', are constrained and/or influenced in their decisions and actions by those higher up in the organisation to which they belong — managers or directors, for example — who may be removed in time and space from the locus of the operator's decisions and actions. As Horlick-Jones (cited in Hood and Jones, 1996: 64) puts it in his discourse on the Inquiry Report:

> ... [T]he evidence suggests that operating procedures ... imposed critical constraints upon the actions of workers .... [T]he 'power of control' of those at the operational end of the corporate body was structurally constrained by those higher up in the management hierarchy .... [I]n organisational settings individual workers do not behave according to the classical models of rational choice. It is not ... a matter of agency arising from individual choice. Rather ... *authority frames ... decisions ... to produce a situation of role requirements and obligations* (my emphasis).

To the extent that the Inquiry Report recognised the — albeit opaque and subtle — influence of management upon staff, the Report could be said to have adopted an holistic or systemic analysis of the aetiology of the *Herald of Free Enterprise* disaster. The authors of the Report looked beyond the *particular* circumstances of the disaster at the 'landscape' of pressures, constraints and requirements that constituted the operational environment of the ferry and its crew. That is, they resisted what might be termed the easier option of dissociating the particular from the general. Instead, they chose to try to understand the disaster in context — the context being a complex, 'messy' reality of competitive and economic pressures, shareholder demands, investment and maintenance decisions, management assumptions, rules and regulations (both corporate and governmental), the frailties of human operators and the unpredictability of the natural environment.

According to Miller (1984: 9), a Royal Air Force (RAF) officer with crash investigation experience, the systems approach to disaster analysis stands in stark contrast to what one might term the 'scientific' or reductionist approach, which, in the early 1980s, was the dominant modus operandi amongst RAF crash investigators. As he explains:

> ... [A]ccident investigators ... usually concentrate on a situation as if it were static and as if it should have been well ordered and in accordance with all the regulations. But we all know that reality is not like that .... *Life is actually far more complex* .... Most of us have been brought up to believe in a scientific approach to problem solving. If we meet a situation which at first we cannot explain, then we adopt reductionist methods to try and produce the explanation. In other words we break down the system into smaller and smaller components until we can identify what we believe to be the cause. We find the weak link and then assume that if we repair it all will be well. If we do this then we neglect the systemic nature of our existence. *Being a system, reality cannot be studied successfully by dividing it into parts, each of which is studied in isolation. Each part is in fact affected by being in the system and is changed if it leaves the system* (my emphasis).

Put simply, says Miller, 'Systemic analysis ... denies that people, objects and events can be treated in isolation. Instead it sees people and objects as being the component parts of a system'. According to Miller the implications of the way in which we understand 'failure' are significant. Thus, on the assumption that an aircraft and its crew are a component part of a larger air transportation system, it is feasible for the system itself, and not just the operator(s), to produce both 'good' and 'bad' outputs. As he explains:

> ... [I]f failure is to be regarded as the output of a *system*, then even though the fault may at first glance appear to lie with the individual himself [sic], the real cause may well be in the system of which he [sic] is but a single part. If this is the case then even our good and experienced pilots are not immune to accidents (1984: 3, my emphasis).

Therefore, as Miller explains, due to the intervention of dynamics over which s/he has no direct or effective control, even 'good and experienced pilots' may find themselves at the locus of, and blamed for a disaster. Perrow (1999) discusses the potential malign influence of external factors on the behaviour of operators in his book *Normal Accidents — Living with High-Risk Technologies*. Focusing on industrial accidents, he makes the following observations on how economic pressures may impact human decision-making and performance:

> Most industrial accidents are attributed to 'operator error' or 'human error' by those who study and seek to prevent such accidents. There is a growing recognition, however, that this is a great oversimplification; worse, it involves blaming the victim. It also suggests an unwitting — or perhaps conscious — class bias; many jobs, for example, require that the operator ignore safety precautions if she is to produce enough to keep her job, but when she is killed or maimed, it is considered her fault (1999: 67).

It is not unthinkable that within the aviation industry, too, meta-economic or customer pressures may find expression in 'corner-cutting' by employees (see Buck (1994: 3-4), above).

Thus systems theory may have a certain utility with regard to understanding the aetiology of disaster. According to Sagan (1993) it can also help to explain—in conjunction with what he calls the 'politics of blame' (1993: 278)—why those at or near the bottom of the organisational hierarchy get most of the blame when something goes wrong. Sagan theorises that operators may suffer a dual victimisation. First, through their (professional and contractual) obligation to operationalise the potentially ill-informed judgements or decisions of managers and/or directors. Secondly, through the subsequent behaviour of those managers and/or directors who, as Sagan points out, may disassociate themselves from the actions of their staff. As he explains in the Conclusion to *The Limits of Safety*:

> A final observation to a theme that ran throughout this book: the politics of blame. When a petrochemical plant explodes, a jumbo jet crashes, or an oil tanker runs aground, accident investigators round up the usual suspects: the control room operator, the pilot, or the captain who committed an error. It is extremely misleading, however, to place such a significant emphasis on 'operator error' as the cause of most accidents. The safety regulations may have been poorly written, but it is easier for plant management to blame the operator, than to accept responsibility itself for writing incomprehensible rules or having poor review procedures. The cockpit switches may have been poorly designed, but it is cheaper to fire the pilot than it is to redesign the control panel. The captain's task may have required absolute perfection, but the ship's owners want the cargo delivered immediately. "Operator error' is an easy classification to make,' as Perrow puts it in his review of dangerous accidents in mines: 'What really is at stake is *an inherently dangerous working situation* where production must be kept moving and risk taking is the price of continued employment' (1993: 278, my emphasis).

The 'working situation' has numerous components. One of the most vital is training. According to a source quoted by Weir (1999: 142) this may be coming under pressure due to commercial considerations:

> Most of the 'big' aviation countries operate anonymous reporting systems whereby pilots, air traffic controllers and other professionals can tell their anecdotes of near-disaster so that others can learn from them. In the UK it is called the Confidential Human-factors Incident Reporting Programme (CHIRP). In October 1997 one pilot for a low-cost airline reported: 'Further pressure has been brought to bear in the training regime, with a reduction in the simulator sessions per year, on a fleet where most co-pilots have less than 1,000 hours' total time, and 50 per cent of captains have less than two years' command experience. Again, 'ticks in boxes' are adhered to but very little real training/development can be achieved in the limited time that is now available'.

In conclusion, pilots, because of such factors as ill-informed design, management decisions or deficient training, may find themselves—or, more precisely, may be placed in—an 'inherently dangerous working situation'. To the extent that this working situation may contain one or more affordances for error, it would seem both unjust and irrational for the pilot to bear the entire burden of blame for a near-miss, accident or disaster. Indeed, in certain cases, it could be suggested that management should bear the major responsibility for the near-miss, accident or disaster. However as Sagan points out, those at the top of an organisation — especially military-type organisations — are rarely held to account. Their power and influence facilitates obfuscation and denial. According to Sagan this dynamic of power constitutes a 'politics of blame'. As he explains:

> The power and self-interest of organisational leaders explains why internal investigations of industrial accidents often find that they were caused by 'operator errors' and were rarely the result of mistaken design or faulty decisions by the senior management. Changing procedures is time-consuming; buying safer equipment is expensive; accepting responsibility for faulty designs is embarrassing. It is often simply easier, as Charles Perrow puts it, to 'blame the victim' rather than the leaders of the organisation. Although this is a widespread phenomenon, hierarchical and tradition-bound military organisations may be particularly prone to blame individuals for errors rather than finding fault with the entire organisation's structures or procedures (1993: 208).

The following section will, through the presentation of a number of case studies, including a military accident, seek to test the views expressed above by those, like Reason, Norman, Perrow and Sagan, who argue for a more systemic or holistic understanding first, of the cause(s) of 'operator error' and secondly of the investigation process and subsequent allocation of blame.

Case Studies

Introduction

A number of case studies will be used to evaluate a range of views on the role of pilot error in near-misses, accidents and disasters. At one end of the spectrum are Shaw (1969) and Tye (1973) who observe that very few aviation disasters are caused by operator error alone. At the other end we find Lufthansa and Boeing— if not the entire aviation industry (Weir, 1999: 140). This evaluation will also allow an assessment of the utility of the systemic or holistic approach to disaster investigation, as advocated by, for example, Miller (1984) and, more recently, Sagan (1993).

The case studies commence with an examination of a military aircraft disaster— the crash in 1963 of a United States Air Force (USAF) Convair (Consolidated Vultee) B-58 Hustler conventional/nuclear bomber over the state of Oklahoma in the United States. While the monograph focuses mainly on commercial aviation disasters, the Oklahoma crash is relevant to the discussion for a number of reasons. First, because it allows an assessment of the role of such factors as pre-production design decisions and operational maintenance regimes and decisions in aircraft accidents. Secondly, because it illustrates the various physical and psychological stressors experienced by pilots (which are, for the most part, generic). Thirdly, because it illustrates the various forms of behaviour of those in positions of authority in the aftermath of disaster and lastly, because it highlights the degree to which authority figures may be influenced by the personal reputation of those directly involved in disaster.

As mentioned above, the format used for this monograph permits the presentation of only a small number of in-depth case studies. In order to provide a comprehensive critique of the opinions described above, the studies feature incidents that are both aetiologically complex and—as with the Convair B-58 case study—multi-dimensional. It is anticipated that this approach will provide the material necessary for an informed commentary on all the assertions made and issues described thus far.

The various case studies will be presented first as factual narratives, with comments being made and conclusions drawn at the very end.

The Convair B-58 accident

Introduction

Consolidated Vultee's B-58 Hustler was developed in response to the USAF's Generalised Bomber Studies of the late 1940s. Designed to fly at Mach 2 at high altitude, the B-58 served just 10 years with the USAF, its demise attributable to the USSR's improved missile defences and the complexity and cost of the aircraft (Tegler, 1999: 36-37). Only 116 Hustlers were built. According to Tegler, although the Hustler's innovative design 'made it a tremendous performer' a number of high-profile accidents 'did not endear it to the air force'. In point of fact, says Tegler, '... for the bulk of its service life, its mishap rate was quite comparable with those of other high-performance aircraft of its day' (1999: 37-40).

The Hustler was of an advanced delta-wing design. It carried a crew of three: pilot, navigator and defensive systems operator (DSO).

The crash—February 14, 1963

The B-58 that crashed on the night of St Valentine's Day, 1963, was being flown by Major John Irving, according to Tegler (1999: 41) a 'vastly experienced' pilot. In 1961 Irving had saved a damaged Hustler by flying it to a desert airstrip where he landed the bomber on a bed of foam. During the emergency, General Powers, head of Strategic Air Command (SAC), had ordered Irving to point the damaged aircraft at a mountain range and then bail out. Major Irving refused the order. Irving was awarded a military honour for saving the aircraft.

The B-58 flown on the night of February 14, 1963 had a design of pitot tube that had a history of malfunction. (The pitot tube is located on the outside of an aircraft near the nose. It is used to provide input for the pilot's airspeed indicator (ASI)). Early model tubes (manufactured by a garbage disposal company called Waste King) had a propensity to freeze up near the ASI. This unfortunate and potentially dangerous defect was attributable to the location of the heating element near the front of the tube, a long way from the cockpit. Any moisture emanating from the heating element tended to freeze near the cockpit, rendering the pilot's ASI inoperable. ASIs provide vital information to the pilot. Without an accurate reading the pilot cannot tell how close his/her aircraft is to stalling. (An aircraft stalls when it is flying so slowly that the wings cease to provide adequate lift. A stalled aircraft will fall from the sky. Recovery is not always possible).

Cognisant of the pitot tube's design defect, the USAF developed a new, more robust pitot system—the 'Melrose System' ('M-system'). Unfortunately, however, this new system had not been fitted to Irving's Hustler due to a shortage of supply. As Irving explains:

The aircraft I had that night had just come out of IRN [inspection and repair as necessary]. At the time the Melrose System was being adopted and a new one was supposed to be installed on this airplane. But the maintenance people didn't have one, so they looked at their logistics manual and found that there was a suitable substitute for the M-system. That happened to be the good old 'Waste King', and they slapped that on the airplane (Irving cited in Tegler, 1999: 43).

True to its history, on the night of February 14, 1963, Irving's 'Waste King' pitot/heater/ASI combination malfunctioned at 35,000 feet. Irving believed that he was flying at close to the speed of sound. In fact, due to the pitot/ASI malfunction, his Hustler was flying at 275 knots (316 mph)—just above the aircraft's stall speed for that altitude. Irving (cited in Tegler, 1999: 44) describes his aircraft's demise thus:

The airplane just quit flying. I was stalled out. I had just lost all airspeed and yet ... my airspeed [ASI] read .92 Mach [close to the speed of sound] .... But the pitot tube was frozen.

After a determined attempt to recover the aircraft, the crew bailed out. All three survived, although Irving was injured.

Following the crash a USAF Accident Investigation Board (AIB) recovered the pitot/ASI combination from Irving's aircraft. According to Irving the AIB, using the recovered device, reproduced the failure *exactly*:

They dug the pitot system up, completely intact. They took it to a wind tunnel and reproduced the atmospheric conditions, temperature and pressure, and the pitot did exactly what it had done to me. It froze up.

According to Irving, despite this finding the USAF did not believe his version of events. As he explains:

With all of the information they had, the pitot system malfunction ... everything ... they charged me with pilot error. The airplane had been grounded for six months because of the pitot system, yet they charged me with pilot error .... [T]he accident investigators hadn't asked me anything. The investigation had gone forward with no input from me (Irving cited in Tegler, 1999: 46).

According to Irving, this decision may have been influenced by the attitude of General Ryan, his senior officer, who seemed surprised that Irving was unable to fly his aircraft within its flight/performance envelope (at night—and hence with few, if any, visual cues—at 35,000 feet) without an ASI. As Irving puts it; '... In my mind, they geared the whole investigation to come up with the answer that Gen. Ryan gave'.

On hearing about the charge of pilot error, Irving, still in hospital, threatened to bring reciprocal charges against both General Ryan and the officer in charge of his aircraft's maintenance programme. Before being discharged Irving received a telephone call from a senior officer in the USAF's Air Force Flying Safety Command. According to Irving (cited in Tegler, 1999: 46) this officer authorised all charges to be dropped, attributing the crash to 'a maintenance and administrative malfunction'. Irving eventually resumed his career with the USAF.

While a detailed theoretical analysis of the allocation of blame within this incident will be offered later, the events outlined above bring immediately to mind the words of Horlick-Jones:

> In disaster causation the action of individuals almost always takes place within organisational settings. This dimension complicates matters considerably, with *the micropolitics of blame* within the organisation distorting both diagnosis of responsibility and processes of learning from past events (cited in Hood and Jones, 1996: 63, my emphasis).

The de Havilland Comet 1 accident

Introduction

The Comet, developed by the British de Havilland company, was the world's first commercial passenger jet. The aircraft had great significance for a gloomy post-war Britain. As Faith (1996a: 53) explains:

> The Comet was conceived in 1946, for the British people one of the gloomiest of the many depressing years immediately after the war. It came to symbolise the dream that Britain could remain at the forefront of aeronautical progress as she had been with radar, the Spitfire, and latterly, with the development of the jet engine.

Wynn Jones (1976: 141) supports Faith's view of the Comet as post-war symbol of national economic and technological re-birth. As he puts it:

> The Comet was more than an airliner in Britain in the early 1950s, it was a national hero .... [I]t is easy to forget the ordinary man's enthusiasm and pride of those heady years ... it ... reflected the country's renewed hopes and aspirations to leadership in the air.

According to de Havilland's test pilot John Cunningham; 'The Comet was a pilot's dream. It was fast, comfortable and smooth. Everyone who flew the jet fell in love with it' (Cunningham cited in Faith, 1996a: 52). The Comet was also popular with the world's airlines. By the end of 1953 even Pan American (Pan-Am), with easy access to and courted by US manufacturers, had ordered the marque (Chant and Batchelor, 1979: 8). Thus it could be said that the Comet placed Britain's aviation industry ahead of that of America. This was no small achievement as the United States had emerged from the Second World War with a strong manufacturing base (Jones, 1983: 578) and its aircraft makers implementing what was to prove a far-sighted strategy. As Williams (1987: 317) explains:

[P]ost-war civil aviation ... was dominated by the USA. There were two main reasons for the commanding position of the American aircraft industry. First and foremost, much of the European industry was in ruins .... [S]econdly the USA formed a very shrewd judgement of the future course of events—especially with regard to the carriage of freight ....

Notwithstanding this early lead, by the 1950s Britain's aviation industry had pulled ahead—at least in terms of passenger jet development. This achievement coincided with (and possibly contributed to) the beginning of a period of relative economic stability and prosperity in Britain (Howlett cited in Johnson, 1994: 320).

The Comet 1 incorporated a number of innovative design features. It was pressurised, so it could fly at altitude 'above the weather'. It also used hydraulic controls. Such controls augment the power that can be applied to an aircraft's control surfaces—essential in such a (relatively) heavy and fast-flying aircraft.

Unfortunately, innovations can have negative as well as positive outcomes. Jet engines, although inherently more reliable than piston engines due to the fact that there is but one major moving part (the spool(s) to which are attached the compressor and turbine blades), are slow to accelerate. As the *Daily Telegraph* explained to its readers after the crash:

... [S]imple jet engines accelerate slowly and do not develop full power until a plane is travelling at high speed and a considerable volume of air is being forced through the engines ... (cited in Hurst, 1976: 46).

During the 1950s there was such concern about this 'power-lag' at low speeds that designers considered fitting rockets to aircraft to provide more thrust during the critical take-off run. Generally a turbojet engine '... becomes more efficient as the speed of the aeroplane increases until it approaches the velocity of the jet'. Turbojets reach peak efficiency at a speed of about 1,100 knots (1,265 mph) (Kermode, 1972: 134-135).

Hydraulic controls can also be problematical as they reduce the level of tactile feedback to the pilot. That is, the pilot's awareness of the impact of his/her inputs to the cockpit's control column on the behaviour of the aircraft's control surfaces (for example, the elevator) is less acute. One might say that hydraulic controls—as mediative devices—bring about a certain sensory detachment from the dynamics of flight, particularly the aircraft's interaction with its physical environment.

Another pertinent characteristic (common to all aircraft) is the way in which the incidence or attitude of an aircraft (the elevation of the nose above the horizon) impacts its drag coefficient. Put simply, the greater the angle of attack (the more one points the nose upwards) the greater the slowing effect of the air through which the aircraft is moving ('drag'). To compensate for this more power is required to maintain the same velocity. This characteristic is useful on landing, when an exaggerated attitude can be used to slow the aircraft (this was a technique used by Hustler pilots when landing the fast-flying B-58 (Tegler, 1999: 39)). On take-off, however, such a characteristic can have unfortunate consequences, with the aircraft, slowed by drag, being unable to attain sufficient speed to leave the ground (unless extra power is applied before the airplane reaches the end of the runway).

The Comet 1 was sensitive to drag. Consequently, de Havilland's engineers had given instructions to pilots not to raise the nose too far during the take-off run. The Comet 1's *Flight Manual* stated:

> At 80 knots [roughly 92 miles per hour] the nose should be raised until the rumble of the nose-wheel ceases. Care should be taken not to overdo this and adopt an exaggerated tail-down attitude with consequent poor acceleration (cited in Hurst, 1976: 47).

The aircraft's sensitivity to drag may have reflected, in part, its use of relatively low-powered centrifugal-type jet engines. As Gilbert (1970: 242) explains:

> The Comet had been first, but something of a venture onto thin ice, for the low power and indifferent efficiency of those early jet engines dictated a small, barely economic [the Comet 1 carried 36 passengers], and lightly constructed airframe.

In fact the Comet 1 was fitted with a design of engine developed concurrently by Britain's Frank Whittle (later Sir Frank Whittle) and Nazi Germany's Hans-Joachim Pabst von Ohain before the Second World War (Chant and Batchelor, 1979: 8). The de Havilland Ghost engines each developed 5,000 lbs of thrust. Given the Comet 1's fully laden weight of 105,000 lbs (Hensser, 1953: 14) this meant that the aircraft had a power to weight ratio of 1:5.25 (that is, every pound of thrust generated was required to lift over five pounds in weight). The de Havilland Trident 1, described by Taylor (1965: 74) as the 'First of the 'second generation' British jetliners' had a power to weight ratio of 1:3.89. The Trident was powered by turbofan engines. Turbofan engines (a synthesis of axial flow turbojet and turbo-prop technologies) have solved the problems experienced during the 1950s. As Williams (1987: 318) explains; '... The turbofan engine [is] a kind of hybrid of the jet and the turboprop, which is relatively quiet and can generate powerful thrust at the low speeds of take-off and landing'.

The crash—October 26, 1952

The British Overseas Airways Corporation (BOAC) Comet 1 that crashed was being flown by Captain R.E. Foote. The aeroplane crashed on take-off from Ciampino Airport near Rome. The crash took place at night in the rain. Under these conditions Foote 'had no visible horizon to guide him'. Unfortunately, during the aircraft's passage down the runway he raised the aircraft's nose slightly above the manufacturer's acceptable maximum angle of incidence. Realising that the Comet was not going to leave the ground, Foote abandoned the take-off. The aircraft overran the runway ending up in some soft ground. There were no serious injuries (Bressey cited in Hurst, 1976: 46-49).

It was originally suspected that an engine had malfunctioned during take-off. However, when Foote's 'error' was discovered both the airframe and engine manufacturers '... showed no mercy on ... Captain Foote'. Despite such mitigating factors as the loss of tactile feedback associated with hydraulic controls and the weather conditions at the time of the accident, Foote was roundly blamed and duly punished. As Bressey (cited in Hurst, 1976: 47) explains:

> In the admittedly difficult circumstances pertaining at the time of Captain Foote's take-off, he should apparently have relied on a 'sixth sense' to prevent the error of 2-3 [degrees] excess incidence which in fact occurred. He was informed that there was no longer a place for him amongst the elite pilots of the Comet team, though he was allowed to retain his command — demoted to the lowliest task for a BOAC pilot — flying York freighters [converted piston engine-powered unpressurised Lancaster bombers].

Some five months after the Ciampino accident another Comet 1 crashed, this time at Karachi during a delivery flight. There were marked similarities between the Karachi and Ciampino crashes. Both happened at night. Both aircraft overran the runway during take-off. Sadly, eleven people died at Karachi (five crew and six technicians). As Bressey (cited in Hurst, 1976: 48) explains, the Karachi crash prompted de Havilland to modify the design of the Comet:

> The manufacturers thereafter carried out a series of tests on the take-off characteristics of the Comet 1 wing, and as a result the wing design was modified, making it much less critical to small errors of judgement in estimating the exact angle to which the nose of the aircraft had to be raised during take-off.

After the Comet no other passenger aircraft design with such a slim margin for error during take-off was to receive approval. The authorities legislated for much larger tolerances (or 'safety-margins') during that most critical stage of getting a large and heavy aircraft safely off the ground. As Faith (1996a: 54) explains:

> As a result of these crashes de Havilland changed the shape of the leading edge of the wing, and as Cunningham says, every single transport aeroplane since then has had to 'demonstrate that it can put its tail on the ground and continue to fly off even with an engine failure'.

Unfortunately, the Comet's troubles continued, with two aircraft being lost to metal fatigue (then an unfamiliar problem to aeronautics engineers). By the time these defects were remedied in 1958, Britain's lead in passenger jet design had vanished. The US manufacturers, especially Boeing and Douglas, were triumphant:

[T]he Boeing 707 airliner [and] the Douglas DC-8 ... made the Comet obsolescent —yet another example of the hazard of being first in a new field. The Boeing 707 and DC-8 had the advantage of bigger capacity, higher speed and longer range (Williams, 1987: 318).

The Ciampino episode inspired a film on the subject of the link between pilot error and design. The 1960 film *Cone of Silence* (British Lion Films) reflected on the fictional Captain Gort who crashed his new commercial jet plane (the Phoenix) by raising the nose too far during take-off. This may be seen as further evidence of the allure of aviation mysteries for the public.

The Munich air disaster

Introduction

On February 6th, 1958, a British European Airways (BEA) piston-engined passenger plane crashed at Munich Airport. Twenty-three people were killed. Seven of the dead played for Manchester United Football Club (MUFC), according to Faith (1996b: 166) '... the greatest [association] football squad ever assembled in Britain'. Perhaps because of this 'celebrity' aspect to the disaster, the crash received, as Bressey (cited in Hurst, 1976: 49) puts it, '... more than the normal amount of publicity'. Faith (1996b: 166) asserts that the crash '... received 'huge publicity''.

The aircraft that crashed was an Airspeed Ambassador. Airspeed was a subsidiary of de Havilland. The Ambassador, like the Comet, had been designed for aerodynamic efficiency. Twenty were built. The aircraft made its last scheduled commercial flight on June 30, 1958, some five months after the Munich disaster (Baldry, 1980: 58).

The crash—February 6, 1958

The weather at Munich Airport during the late afternoon of February 6th was not good. 'Soft unfreezing snow' was falling, leaving the runway 'slush covered'. The Ambassador's captain, James Thain, had made two runs in his aircraft, after which he had told a BEA engineer that he was unhappy with the situation. On the third run Captain Thain aborted the take-off 'at a rather late stage'. Unable to stop the aircraft on the runway, the Ambassador '... careered off the end at considerable speed, veered to one side, struck a house and some trees, broke up and caught fire' (Bressey cited in Hurst, 1976: 49-53).

There was immediate speculation as to the cause(s) of the disaster—some of it from 'official' quarters. Thus on February 8th, BEA's Chief Executive stated that the accident '... was turned into a major disaster by the house situated 300 yards from the end of the ... runway'. (In fact the location of the house conformed to contemporary planning regulations). Following the Chief Executive's statement, the West German Traffic and Transport Ministry announced that it believed ice on the aircraft's wings to be the 'probable reason' for the Ambassador's failure to leave the ground. (Ice deposits reduce the aerodynamic efficiency of aerofoils, thereby reducing lift). On the day of the crash de-icing sprays were not being employed by the airport authority (Bressey cited in Hurst, 1976: 49-53).

In 1959 the German Ministry of Transport published its official report into the disaster. It asserted the 'decisive cause' of the accident to be ice on the Ambassador's wings — despite the fact that, as Bressey (cited in Hurst, 1976: 50) explains; '[E]vidence as to the actual existence of this ice was, at best, conflicting'. Because the Captain is responsible for establishing that his aircraft is airworthy (achieved partly through the 'walk around', where the flight crew physically survey the aircraft before departure) a concomitant of this finding was that the Germans held Captain Thain responsible for the disaster. Thain allegedly did not check for ice.

The report made only 'passing reference' to slush on the runway, and, according to Bressey (cited in Hurst, 1976: 51), when the issue was discussed '... [a] completely unfounded statement was made':

> All experience goes to show ... that it may be assumed that take-offs can be made with nose-wheel aircraft without danger up to a slush depth of at least 5 cm [2 inches] (Commission of Enquiry cited in Hurst, 1976: 51).

It should be noted, however, that at the time of the German Commission of Inquiry there was little scientific knowledge about the effect of slush on aircraft take-off performance.

Unhappy with the Commission of Inquiry report, Captain Thain lobbied for and obtained a British Court of Inquiry. Chaired by Mr Edgar Fay QC the Court of Inquiry published its report in August 1960. According to Bressey, the 'Fay Commission' '... appeared to agree ... that the major cause of the accident was wing icing'. However, Bressey asserts that the Fay Commission had been given ' ... extremely limited terms of reference in order to avoid political embarrassments with Germany' (cited in Hurst, 1976: 51). The following year the UK's Ministry of Aviation published an *Information Circular* that '... warned pilots that only half an inch [1.3 cm] of slush on the runway would increase the take-off run by 40 per cent for nose-wheel aircraft ...' (Bressey cited in Hurst, 1976: 51). (The Ambassador had a three-point undercarriage). The Ministry of Aviation continued to investigate the effects of slush on take-off performance, even using an Ambassador in their trials. The results were shared with the Germans. Eventually, in 1966 the German authorities published a report which gave some credibility to the British theory, but, as Bressey (cited in Hurst, 1976: 52) explains; '... [I]t still maintained that 'Wing icing was an essential cause of the accident''. In 1968 the

UK's Board of Trade reopened the Inquiry. Mr Edgar Fay, who had chaired the first Inquiry, presented his conclusions in 1969. The report exonerated Captain Thain. As Bressey (cited in Hurst, 1976: 52-53) explains:

> The report [stated that] in the light of present knowledge, it was their opinion that slush on the runway was the prime cause of the accident. However, so little had been made public on this phenomenon 11 years ago that Captain Thain could not have been expected to take this into account. Finally: 'In accordance with our terms of reference we therefore report that in our opinion blame for the accident is not to be imputed to Captain Thain'.

According to Bressey (cited in Hurst, 1976: 53) despite this unequivocal conclusion '... the German authorities ... persistently refused to accept the findings of the Second Fay Commission, without producing valid reasons for this refusal'. Captain Thain died in 1975 aged 53. He had been suffering from a heart complaint.

The Air Canada 'Super 63' disaster

Introduction

On Sunday July 5th 1970 an Air Canada DC-8 Super 63 Model crashed at Toronto. Everybody perished. The 100 passengers and 9 crew died when the aircraft, on fire, crashed into a field five miles north of the airport.

The DC-8 Super 63 was a large aircraft by the standards of the day. Powered by four 19,000 lbs-thrust Pratt and Whitney turbofans the aircraft could accommodate up to 259 passengers. It carried up to 11 crew. The DC-8 design was a 'narrowbody' in contrast to the new 'wide-bodied' Boeing 747. The latter, just coming into service, could carry up to 500 passengers.

The DC-8's engines were located beneath the wing on pylons. As with other aircraft the DC-8 had 'wet' wings, meaning that they contained fuel tanks. The DC-8 had lift-dumping devices called 'spoilers'. These extensible plates were controlled by a lever in the cockpit. Lifting the lever 'armed' the spoilers. Pulling it back deployed the spoilers. Deployment of the DC-8's spoilers in flight brought about rapid height loss. A common technique was to arm the spoilers when the aircraft was near the ground, thereby allowing automatic deployment when the wheels of the aircraft touched the runway (Gero, 2000: 95). According to Gero (2000: 96); 'Air Canada procedures ... dictated that the system be armed for automatic deployment at an altitude of 1,000 feet'.

The crash—July 5, 1970

As Brookes (1996: 44) explains, during the DC-8's approach to Toronto the crew '... agreed that [the] lift-spoiling devices would be armed during the flare [that is, just seconds before touch-down] allowing them to extend automatically once the wheels made contact with the runway'. At an altitude of sixty feet the First Officer not only lifted the lever, but also pulled it back. The spoilers extended causing the DC-8 to sink rapidly towards the ground. On hitting the runway the starboard outer engine and its mounting fell off the wing. This trauma severed some electrical wiring, which then ignited leaking fuel. The Captain decided to fly another circuit. He successfully retracted both the landing gear and spoilers. Three explosions followed, the third causing the starboard outer wing to fall off. The aircraft hit the ground at 250 mph, having fallen from 3,000 feet.

As Brookes (1996: 45) explains, the crew violated company policy on the operation of the spoilers. Air Canada's policy was that the spoilers should be armed at 1,000 feet. Clearly they were armed — and, most disastrously, deployed — at a much lower altitude.

The primary cause of the disaster was the deployment of the spoilers in close proximity to the ground. However, the Board of Inquiry highlighted a number of *secondary* causes. First, a deployment mechanism (the single lever) that provided for both the arming and extension of the spoilers — albeit via different 'actions' on the lever. The Board were critical of the fact that two crucial or 'primary' functions had been combined in a single actuating device. As Brookes (1996: 45) explains:

> [The] Board of Inquiry blamed the disaster as much on faulty design as on human error. It argued that while a single actuating mechanism might be acceptable to perform different secondary tasks such as heating or ventilation, this arrangement was completely unacceptable for something as crucial as lift spoilers. The Board believed that, at the very least, the activating lever should have been fitted with some guard or gate.

The possibility of crew error was compounded by an operational factor, namely the impression given by the DC-8's instruction manuals that, as Brookes (1996: 45) puts it; '... [T]here was a mechanism built-in to prevent the spoilers from in-flight extension'. Brookes goes on to assert that; 'Consequently, Air Canada training staff did not realise that ... inadvertent deployment ... was possible. This ensured that the airline's pilots were never made aware of the potential threat to flight safety'.

The Board of Inquiry also criticised the structural integrity of the DC-8, specifically the design of the turbofan/pylon installation, of the wing tanks and of the electrical wiring. Initially, remedial measures consisted of a Federal Aviation Administration Directive ordering the installation of placards warning against the in-flight deployment of spoilers. Then, after a non-fatal accident about three years later the FAA produced a second Directive '... requiring that all DC-8s be fitted with spoiler-locking mechanisms to prevent such an occurrence' (Brookes, 1996: 45).

The DC-8 completed its first commercial flight in September 1959. As Brookes explains; '[The] safety device, which could have been incorporated at the initial design stage for minimal cost, only became mandatory some 12 years after the DC-8 first flew'.

While, following Brookes's argument, Douglas failed to exercise foresight in the matter of the design of the spoiler deployment mechanism, they were not so remiss in other (commercially-oriented) design decisions. Thus, from the outset, Douglas made a series of crucial decisions that ensured that their new airframe had significant 'growth potential'. As Chant and Batchelor (1979: 12) explain:

> By 1955 Douglas felt that sales fall-off indicated that it was time to put new life into the basic DC-8 airframe to meet the demands of the booming travel industry. Such growth has been allowed for in the original design of the DC-8, in the form of 'fuselage plugs' that could be inserted in front of and behind the wings .... The basic DC-8 design had made special provision for long undercarriage legs ... and an upswept rear fuselage to allow fuselage plugs to be inserted without causing any rotational problems to the aircraft at take-off [that is, without causing the tail to hit the runway as the nose was lifted].

This referencing or association of the initial design to/with forecasts of market growth and customer-preference gave the DC-8 design an advantage over Boeing's equivalent, the 707, which had limited 'stretch' potential. So successful were Douglas's sales team that, despite the fact that the DC-8 was announced more than twelve months after the 707 prototype had been rolled out by Boeing, by the end of 1955 Douglas had 98 orders for the DC-8 to Boeing's 73 orders for the 707. Douglas did not build a prototype, choosing instead to put the aircraft immediately into production. The company '... implemented an accelerated flight trial programme .... [T]ime between first flight and certification date was fifteen months for the DC-8 compared with fifty-one months for the 707-120 ...' (Francillon, 1979: 580-583). It might be asked whether this accelerated programme reduced the design effort for certain components (perhaps the spoiler lever?) below acceptable minima.

The Trident disaster

Introduction

On Sunday 18th of June, 1972, a Hawker Siddeley Trident 1C passenger jet crashed at Staines, shortly after taking off from Heathrow Airport just west of London. Everyone on board was killed. After it had stalled, the aircraft, belonging to British European Airways (BEA), dropped from the sky. It 'pancaked' onto its belly in a field.

The Trident aircraft had been designed, like the ill-fated Comet, by de Havilland (as the DH-121). Like the Comet, the Trident broke new ground. It was the world's first three-engined jet airliner. Additionally, as Chant and Batchelor (1979: 24) explain; '... [It] played a very important part in the development of fully automatic blind landing, a key factor in modern aviation safety ...'. To this end it was fitted with the Smith's Flight Control System. The aircraft had a novel configuration, having all three of its Rolls Royce Spey turbofans located at the rear of the aircraft. The great benefit of this configuration is the contribution it makes to the aircraft's fuel efficiency—and therefore the aircraft's overall economy of operation. Putting the engines at the rear of an aircraft 'cleans up' the wing. Drag is reduced. The main aerofoil operates more efficiently (Jackson, 1978: 502). Other aircraft have made productive use of this economy-enhancing design feature, for example the Fokker F28, Douglas DC-9, Boeing 727, BAC 1-11, and (Soviet) Tupolev Tu-134. All these aircraft were developed as fast, economical short to medium haul commercial transports. Despite the promising economics of this design, however, it does have one fundamental aerodynamic weakness: at certain 'angles of attack' (that is, when the nose of the aircraft is pointing upwards at a certain angle from the horizontal) the wings can interrupt the flow of air to the elevator (in rear-engined aircraft mounted atop the fin to make room for the engines). Should this occur when the aircraft is stalled, recovery can become problematical. This inability to recover a stalled 'T-tail' aircraft is known as a 'deep stall'. On June 3, 1966 a production Trident on its maiden flight crashed in Norfolk, England killing both test pilots and the two other crew members. The aircraft had entered a deep stall from which the pilots (who, as test pilots, would have been highly skilled) were unable to recover (Jackson, 1978: 502).

Paradoxically, the problem of deep stall does not apply to earlier, more conventional designs of aircraft. As Edwards (1974: 79) explains:

> Earlier generations of aircraft were usually found to have very good stalling characteristics. The pilot had ample warning of an approaching stall from the 'buffeting' which he could feel both through the control wheel and in the 'seat of his pants', and if he allowed the stall to develop the aircraft's nose would eventually pitch safely downwards. But some of the more recent types of jetliner with rear-mounted engines and a T-shaped tail may instead pitch nose-up into a 'super-stall', from which recovery is much more difficult since the tailplane is blanketed by the turbulent airflow of the wings. Before such an aircraft can be certificated for airline operation the authorities insist that it shall be equipped with virtually foolproof [sic] anti-stall devices. These will probably include a 'stick-shaker' to give an unmistakable pre-stall warning to the pilot [by vibrating the control column], and a 'stick-pusher', which initiates the recovery automatically by pushing the wheel forward [that is, it pushes the control column forward, which lowers the nose, thereby stabilising the airflow over the wings].

Hawker Siddeley (who bought de Havilland) fitted both a stick-shaker and stick-pusher to the Trident. The designers gave the crew the option of cancelling the stick-pusher (using a lever on the flight deck pedestal). The aircraft had other important design features. For example, it had both flaps and 'droops', extending respectively from the trailing and leading edges of the wing. These devices increase the surface area of the wing, providing increased lift at low speeds. (Droops are attached to and pivot about the underside of the leading edge of the wing. The Trident was the first British commercial airplane to be fitted with droops). The loss of these 'high-lift' devices at low speeds may cause the aircraft to stall. Premature retraction of the droops increases the Trident's stall speed by about 30 knots (35 mph). The flaps and droops were operated by levers. These levers, mounted on a quadrant, operated independently. Thus there was nothing to prevent the flaps and droops being retracted simultaneously (which, without an increase in power and forward airspeed would result in a potentially dangerous loss of lift for a given airspeed). Should the Trident be in danger of stalling, a number of sensory warnings would be activated. As Scott (cited in Faith, 1996a: 174) explains:

> First an amber light would have gone on in front of each pilot, as well as a 'droops out of position' warning light in the front of the central pedestal, followed by a stick-shaker ....

According to Faith (1996a: 174) the stall-warning system '... had previously proved unreliable, resulting in several false alarms'.

British European Airways placed an initial order for 24 Tridents. According to Jackson (1978: 501); 'Confidence in the manufacturers ... was such that no prototype was ordered and the aircraft went into quantity production straight from the drawing board'. Overall the Trident was not a great commercial success. Some 117 Tridents were built. By 1978 over 1,300 Boeing 727s (an aircraft of similar design) had been built and sold.

The crash—June 18, 1972.

The weather at the time of the crash was not good—blustery with low cloud and rain. The cloudbase stood at 1,000 feet. The Trident was Captained by Mr Stanley Key, a 'highly experienced' (Job, 1994: 88) fifty-one-year-old pilot. The flight crew also consisted of a twenty-two-year-old First Officer with forty hours experience as co-pilot, and a more experienced (twenty-four-year-old) 'monitoring pilot', located on a seat behind the centre pedestal. There was a fourth person on the flight deck — a Vickers Vanguard (an older turboprop aircraft design) Captain in transit to Brussels. He also had Trident experience.

Industrial relations within BEA at this time were in some difficulty. As Job (1994: 88) explains:

> The London base of British European Airways was a far from happy workplace in June 1972. A long-standing industrial dispute between the airline and the British Airline Pilots' Association [BALPA] over working conditions and rates of pay was straining relationships, not only between management and flight crew, but

between individual pilots themselves ... a majority were in favour of strike action, but there were a number, in particular senior pilots of 'the old school' who considered such conduct unbefitting and unprofessional.

According to Job, Key belonged to the 'old school'. As Job (1994: 88) explains, Key '... was vehemently opposed to the strike action mooted by the more militant BEA pilots'. Further, says Job, 'His stand on the issue had earnt him considerable criticism, if not vilification, among more junior flight crew ...'. While waiting in BEA's crew room prior to boarding the Trident, Captain Key had engaged in a furious argument with another pilot. This argument had been witnessed by his twenty-two-year-old First Officer.

The Trident took off from runway 27R (which pointed due west). Sixty-three seconds after commencing its take-off run the crew engaged the autopilot. Seventy-four seconds after starting its take-off roll the aircraft began banking to the left. At ninety-three seconds flaps were selected 'up' and the crew commenced the 'noise abatement' procedure (by reducing engine thrust). At the time 'climb power' was deselected in favour of the noise abatement thrust setting; the aircraft was flying at about 168 knots (193 mph). By the time the aircraft made its final radio transmission some 108 seconds after starting to roll, it was flying at a speed of 157 knots (181 mph). According to Job (1994: 93) this was '20 knots [23 mph] below noise-abatement climb speed' (of 177 knots (204 mph)). At 114 seconds, at an altitude of about 1,770 feet, the droops were 'unaccountably' selected 'up'. According to Job (1994: 91) this put the aircraft '... into an incipient aerodynamic stall ... triggering both the stick shaker stall warning and the stick pusher stall recovery system'. At 116 seconds and a height of about 1,790 feet the autopilot, in response to the activation of the stall-warning and recovery system, disengaged. The aircraft recovered for a short while, then stalled again. Two stick pushes occurred. During the third, a crew member disengaged (or 'dumped') the stick-pusher. The Trident was now at an altitude of 1,360 feet. According to Job (1994: 94); 'It was a fatal mistake'. The Trident entered a deep stall, during which the aircraft adopted a nose-up attitude of 31.3 degrees. Recovery was now almost impossible. The aircraft crashed close to the A30 Staines by-pass road. The rapid vertical deceleration at the moment of impact killed all but one on board. The survivor died later. There was no fire, enabling careful examination of the aircraft's structure and systems. As Job explains:

> Despite a most searching examination, no defect or evidence of malfunction was found in the aircraft or its various systems and it was clear that the Trident was fully serviceable in every way up to the moment of its impact with the ground .... [T]he only factor apparently preventing it from flying normally was its almost complete lack of airspeed (1994: 91).

The aircraft's 'almost complete lack of airspeed' was the end product of a flight during which, according to Job (1994: 92) the Trident '... was flown [in a manner that] differed disturbingly from standard BEA practice—it had consistently failed to achieve the appropriate airspeeds for the various phases of flight'.

The absence of a post-impact fire enabled autopsies to be conducted on the crew. Captain Key was found to have been suffering from atherosclerosis (a narrowing of the arteries). Experts suggested that Captain Key had suffered a haemorrhage not more than 120 minutes before the crash. This health problem could have prevented Captain Key from performing at peak efficiency. As Job puts it:

> The symptoms of the internal haemorrhage ... could range from a slight indigestion-like pain in the chest to collapse and unconsciousness. At the very least, it was considered they would have caused some 'disturbance of thought processes' (1994: 94).

According to Job (1994: 93-94) dire though its situation was, the aircraft could have been recovered from its stalled state by either a) increasing air speed by only 10 knots (12 mph), b) extending the droops or c) applying pressure to the control column after the operation of the stick-pusher to maintain forward speed.

According to Bressey (cited in Hurst, 1976: 61); 'Like so many crashes, this one owed much to a combination of circumstances, none of which would necessarily have proved fatal in isolation'. Bressey (cited in Hurst, 1976: 61) says that these circumstances included:

a)   the possibility that Captain Key had suffered a cardiac arrest;

b)   the rostering of two relatively inexperienced pilots to fly with the experienced Key;

c)   in Bressey's own words; 'The necessity to adhere to a complicated 'noise abatement' flight profile, which, although officially 'safe', nevertheless reduced the performance safety margin below the optimum'; and

d)   '... factors which permitted the retraction of the leading-edge droop during the critical phase of flight'.

According to Gero (2000: 108) there may also have been a spatial or situational disorientation dimension to the disaster. As he puts it; 'At the most crucial time ... the aircraft was operating in cloud, and the crew had no outside visual reference'. Thus there are similarities between the flight-deck situational awareness dynamics of this disaster and the earlier B-58 and Comet disasters.

The crash report attributed the disaster to twelve causes — five 'immediate' and seven 'underlying'. The *most immediate* cause was the premature retraction of the droops by a crewmember. The report listed the last underlying cause as; 'Lack of any *mechanism* to prevent retraction of the droops at too low a speed after flap retraction' (cited in Hurst, 1976: 61, my emphasis). Scott (cited in Faith, 1996a:174) postulates that the general confusion within and without the flight deck may have led to the crew misreading their airspeed:

> There was a whole cacophony of sound and light going on all at once [the various sensory warning cues referred to above], totally unexpectedly, as well as the aircraft buffeting around in the weather conditions at the time. It's quite understandable how the crew didn't recognise their predicament.

Of course, it is also possible that the two co-pilots did recognise that their aircraft was flying too slowly, but deferred to their Captain. If they had no inkling as to his state of health, it could be argued that they would have less reason to challenge his judgement and actions. Had cockpit voice recorders (CVRs) been fitted to British commercial aircraft like the Trident, then the aetiology of the disaster might have been revealed. Unfortunately, as Weir (1999: 221) explains, both BOAC and BEA had resisted the installation of CVRs.

Finally, if, as Bressey (1976) alleges, the crew did indeed defer to their Captain, it should be noted that even with such contemporary innovations as 'crew resource management' (CRM)—a 'psychology of the cockpit'—deference is still to be found on the flight deck. Consider, for example, the following report by Jones (2000):

> [O]n June 6, 1998 ... [t]wo aircraft ... descended too low, rushed their approach and narrowly missed the runway when landing at Ronaldsway Airport .... In the first case ... the first officer had worries about the approach but was 'reluctant' to question his commander because he had been with [the airline] only a short time.

The MALEV mystery

Introduction

MALEV was the national airline of Hungary during the Soviet era. Given Hungary's close association with the USSR, the national airline operated Soviet-designed and built aircraft. These ranged in the 1960s and 1970s from, as Barlay (1997: 38) puts it, 'unsophisticated Russian turboprops' to relatively modern jet transports like the Tupolev-134 (TU-134). The TU-134 was designed in the 1960s and made its first commercial flight with Aeroflot (the Soviet Union's national airline) on September 9th, 1967. Numerous eastern bloc airlines bought the type. MALEV also operated the larger, three-engined Tupolev-154.

The design of the TU-134 followed the 1960s fashion for placing the engines at the rear of the aircraft. According to Chant and Batchelor (1979: 20); 'This left the wing free to perform its aerodynamic functions, reduced noise and vibration levels and gave good slow speed handling characteristics'. (Of course, there was always the residual risk of a deep stall). The aircraft was powered by two turbofan engines. Turbofans strike a compromise between the good low-speed performance characteristics of the turboprop and good high-speed performance characteristics of the pure jet engine (Huenecke, 1997: 9). (Concorde, for example, uses turbojets. The aircraft cruises at just over twice the speed of sound at high altitude). It carried a crew of five and, in its TU-134A incarnation, up to 80 passengers. By 1979 some 350 had been produced. MALEV received the TU-134 from 1969 onwards. The airline also operated the TU-154, which could carry up to 167 passengers. Like the TU-134, the TU-154 had its engines located at the rear of the aircraft, two on either side of the fuselage and one under the tail. This was the same configuration as that employed in the Trident.

The 1970s — a decade of turmoil

In the 1970s MALEV suffered a number of major crashes. In 1971 a TU-134 crashed at Kiev during its approach, killing 49. In 1975 a TU-154 crashed at Beirut during its approach killing 60. In 1977 a TU-134 crashed at Urziceni in Romania during its approach killing 29 (Bordoni, 1997: 132-192). According to a senior MALEV employee; 'Accusations were flying in every direction, above all against the pilots ...' (Matyasy cited in Barlay, 1997: 37). Eventually a government committee investigated the airline's safety record. Its findings are interpreted here by the same employee:

> Eventually, the root of all problems was found: management. We had incessant changes of chief executive and chief engineer — most of them professional soldiers, chosen for political reliability, who demanded more parade ground discipline than aeronautical professionalism .... [O]ur elderly maintenance staff had mostly been trained as car mechanics, and sent to the Soviet Union for two or three months of theoretical but hardly any practical conversion courses before assigning them to work on aeroplanes! They just about managed to keep our unsophisticated Russian turboprops flying. But in 1969 MALEV received the TU-134s, the first jets to replace our Ilyushins [turboprop-powered passenger aircraft] .... [T]he management ... was also quite happy to take on retired fighter pilots rather than train new professionals (Matyasy cited in Barlay, 1997: 38).

Having identified the problem, remedial measures were taken. Professional managers and aircraft mechanics were hired and standards of operation and maintenance were improved. According to Matyasy (cited in Barlay, 1997: 38); '... [A]s if by waving a magic wand the airline was safe again'. Following the collapse of the Soviet Union at the end of the 1980s MALEV had access to Western aircraft manufacturers. However, according to Matyasy, MALEV resisted the temptation to introduce more modern equipment immediately. Rather, the airline took stock and resolved to go through another process of readjustment in preparation for a more modern fleet. One might conclude that the Hungarian airline had learned a lesson. By 1997 the airline was operating Boeings and Fokkers, in addition to its original Russian-built jets (Taylor, 1997).

Conclusion

The one military and five civil accidents/disasters described above constitute the case study component of the monograph. They were chosen to illustrate the potentially complex and unpredictable aetiology of aviation accidents and disasters. In the Discussion and Conclusions section the case studies will be reviewed in light of the work of academics like Lee and Shapero, Shaw and Tye, Toft, Horlick-Jones, Reason, Norman and Sagan as well as the numerous 'non-academic' commentators cited above, like Faith, Barlay and Bressey. The case studies will be analysed in sequence. Final (tentative) conclusions will be drawn at the end of the monograph.

Discussion and Conclusions

The B-58 Hustler accident

Clearly the *immediate* cause of the crash of the B-58 on the night of February 14, 1963, was Major Irving's failure to maintain an appropriate velocity. There were, however, mitigating circumstances. The ASI had malfunctioned at an altitude where the pilot was required to fly his aircraft within a relatively narrow 'speed band'. (At ground-level the speed range between flying so fast that the aircraft encounters 'high speed buffet' and flying so slowly as to risk a stall is quite large. At altitude this 'safety envelope' gradually reduces to the point where the pilot has to concentrate on flying within a relatively narrow speed band (Evans, 1997: 58-60)). Flying at night with few, if any, environmental cues, Major Irving failed to fly his aircraft within the required speed band. Of course, it could be argued that such a skilled flyer should have been able to fly the aircraft safely 'by the seat of his pants' — without either an ASI or visual references. As Tegler (1999: 46) explains, this view underpinned General Ryan's criticisms of Major Irving. When informed by a fellow officer that Irving had failed to maintain an adequate airspeed without his ASI, Ryan is alleged to have retorted; 'You mean we have pilots who can't fly without an airspeed indicator?' It may well be the case that such a skilled aviator should have been able to fly the B-58 without an ASI. However, working on the assumption that this, for the reasons outlined above, would have been difficult, it is worth deconstructing the disaster's convoluted aetiology further.

As mentioned in the case study, the design of the pitot/heater/ASI assembly was defective. Although the USAF recognised and attempted to remedy the defect it could be argued that the February 14th crash had its origins in both the USAF's original bomber specification and/or design decisions made by Waste King and/or the integrity and comprehensiveness of both Convair's and the USAF's acceptance testing of the new and complex bomber. Following Reason's discourse, the original Waste King pitot/heater/ASI assembly constituted a 'resident pathogen' replicated within the hull of each and every B-58—a 'latent defect' that, under 'facilitative' circumstances, would transmutate into an active (and potentially lethal) defect.

Following Horlick-Jones's (cited in Hood and Jones, 1996: 63) discourse on the 'micropolitics of blame' and Sagan's (1993: 208) allegations of blamism within 'hierarchical and tradition-bound military-type organisations' it could be said that Major Irving was, in some degree, victimised by a military establishment seeking to distance itself from disaster. To the extent that the military establishment was represented by General Ryan, the USAF refused to accept *any* responsibility for the events of February 14th—even though it had knowingly allowed Irving to fly an aircraft with a potential serious defect (the Hustler had been grounded for six months because of the defective Waste King pitot/heater/ASI assembly). It was only after the intervention of another general (from the service's Flying Safety Command) that the USAF admitted the underlying causes of the disaster — '... maintenance and administrative malfunction'.

The initial mode of the USAF's disaster investigation is also worthy of comment. Despite his first-hand knowledge, no-one thought of interrogating Irving (although a written submission was considered). Given that questioning might have generated a more vivid and accurate picture of the disaster it could be said that the AIB's aloofness was short-sighted. Bressey (cited in Hurst, 1976: 45) makes the following comment on the importance of assimilating all possible sources of evidence into an investigation:

> True accountability ... must be a matter for the patient ... establishment of fact; for the questioning of witnesses, the collation and analysis of evidence .... It is clear ... that no judgement which ... leaves any vital submission unexplored can be regarded as being in any way valid.

To the extent that the USAF's initial investigation did not assimilate 'all possible sources of evidence' it could be said that the AIB's inquiry was invalid and the resulting ascription of blame misplaced.

Pitot/ASI malfunctions still occur. In February 1996 a modern commercial aircraft crashed in the Dominican Republic killing 189 passengers and crew. The aircraft's autopilot, fed by the Captain's pitot/ASI assembly, had reduced the aircraft's speed to just above stall speed. The Captain's ASI read 350 knots (403 mph); the First Officer's ASI read about 200 knots (230 mph). (This was possibly the correct speed at the time). The stickshaker activated. The crew fought with the aircraft, but in vain. The crash investigators cited the 'probable cause' of the disaster as; 'The crew's failure to recognise the activation of the stickshaker as a warning of imminent entrance to the stall ...'. It is possible that the Captain's erroneous ASI reading was caused by an obstructed pitot tube (Aviation Safety Network, 1996). Thus, as with the Hustler disaster, the pilots have been held responsible for an event that probably originated in equipment malfunction. Unfortunately, unlike Major Irving, they are not alive to mount a defence.

The Comet 1 accident

There is no doubt that the Comet 1 accident at Ciampino was roundly blamed on the pilot. Certainly the *immediate* cause of the overrun was Captain Foote's misjudgement of the aircraft's angle of attack while on the runway. In attempting to fly the aircraft off at an 'excess incidence' of 2-3 degrees the low-powered aircraft was unable to leave the ground. This potential problem had been recognised by de Havilland who had issued explicit instructions to pilots not to raise the aircraft's nose too far during the take-off run. Other measures had also been taken by the manufacturers. The aircraft's structure had been 'optimised'. That is, in an effort to improve the power-to-weight ratio, weight had been saved wherever possible. (Given the design's later problem with catastrophic metal fatigue, it might be said that the structure had been over-optimised).

However, as with the Hustler crash, the accident had a number of underlying, or what one might term 'long-lead' causes. Most profound of these was the necessity to stretch the technology of the day to its absolute limits. Had this not been done, de Havilland could never have manufactured the Comet as an economic proposition for the world's airlines. Thus to compensate for the low power and inefficiency of the early jet engines, the airframe was compact and lightweight. De Havilland, realising that this was still not enough to compensate for the power-lag of the aircraft's four Ghost jet engines, required pilots to fly the aircraft off the runway within a pre-specified safety envelope. Given Foote's inability to fly his aircraft within this safety envelope at Ciampino it might be argued that de Havilland's stipulation placed unrealistic expectations on pilots. If Foote, an experienced and skilled pilot, could not do it what chance might others have? The retort might be that if de Havilland's test pilots could fly the aircraft off safely, then any pilot should be able to meet the manufacturer's operational requirement. However, as *Flight International* (1999) points out, 'line pilots' may not be as skilled, or have as much leeway in terms of time as test pilots. There were other mitigating circumstances, too. Foote was taking off at night in bad weather. Consequently, there were few visual references to help him keep the aircraft within its relatively narrow safety envelope for the take-off run. Also, the aircraft was fitted with the new hydraulic controls which, as mentioned earlier, would not provide the pilot with as much sensory feedback as manual controls.

Thus a systemic or holistic analysis would appear to suggest that de Havilland should bear at least some of the responsibility for the Ciampino crash. It could be argued that the manufacturer, through its design decisions and operational requirements, had placed unrealistic expectations on Captain Foote *under the circumstances obtaining at Ciampino on the night of the accident.*

It could be said, however, that to discover the ultimate cause of the accident one should cast the net even wider. As Reason (1990: 188) puts it; '... causal chains [may be] traced backwards in time 'raising the possibility of uncovering '... a combinatorial explosion of possible root causes [of disaster]'. Thus, to the extent that de Havilland were responding to the *politico-economic* aspirations of the British people (articulated through the Government of the day) then perhaps ultimate responsibility for the crash lay with the public at large. Post-war Britain was a bleak place. Austerity was the watchword. The public were looking for a sign that their lot would improve. Partly this could be achieved by economic development (or, in light of wartime losses, re-development). Therefore it could be argued that de Havilland were, in some degree, simply responding to a national ambition: to be first in the newest form of aeronautical endeavour—jet-powered commercial aviation. Unfortunately for de Havilland, the technologies of the day were not as advanced as the politico-economic ambitions of the British people and their Government.

The Munich air disaster

The first inquiries into the Munich disaster cited ice on the wings as the main cause of the crash. Given that it was the Captain's duty to ensure that his aircraft was fit to fly, and the fact that Captain Thain did not check for ice, the implication was that Thain's omission contributed to the disaster.

However, as mentioned above, the wing icing theory may have been a 'red herring'. At the time of the disaster the runway was covered with slush. Slush has the effect of retarding an aircraft's speed during the take off run. As this scientific fact was not known at the time of the first inquiries, it was perhaps inevitable that attention focused on the possibility of wing-icing. Nevertheless, given this 'state of ignorance' about the effects of slush on momentum, the German statement that take-offs could be made '... without danger up to a slush depth of at least 5 cm' was precipitous. It could be suggested that, given the degree of scientific uncertainty, the German authorities should have been rather more equivocal.

Following Sagan's (1993: 278) and Horlick-Jones's (cited in Hood and Jones, 1996: 63) comments on the 'politics of blame' it could be argued that both the German and British inquiries 'scapegoated' Thain as a means of drawing attention away from weaknesses both within the air transport system and its knowledge base. Thus the Germans dealt with the issue of scientific uncertainty in the matter of the retarding effect of slush by unequivocally discounting slush as a factor in the disaster. Through their statements they gave the impression that they were in possession of knowledge that, at the end of the 1950s, did not exist. This might be called disingenuous. Furthermore, neither party considered the possibility that Thain might (like all Captains) have felt some pressure to fulfil his operating schedule. As Buck (1994: 3-4) puts it; '... passengers ... add pressure unknowingly with their gripes and demands to make schedules for their connections ...'. Lastly, despite the bad weather at the time of the disaster, the Munich Airport authorities were not spraying aircraft with de-icer. Given the fact that the German (and first British) inquiries concluded ice formation to be the primary cause of the disaster, this fact would seem to expose the German authorities to a charge of negligence in the matter of ground handling operations.

In conclusion, while the three primary causes of the Munich air disaster were (assuming ice to have been at least a partial factor) Captain Thain's failure to check for wing ice, his decision to make a third attempt to take off and his late abort of the take-off, a systemic or holistic analysis reveals several potential underlying causes: poor weather; a lack of scientific knowledge about the effects of slush on take-off performance; commercial pressures; the presence of a house close to the runway; and, assuming wing ice to have been a factor, the absence of airport de-icing. Following Perrow's (1999: 249) argument, these underlying causes could, in combination, be said to have constituted '... an inherently dangerous working situation' for Captain Thain.

The Air Canada DC-8 disaster

Without doubt the primary cause of this disaster was the First Officer's in-flight deployment of the aircraft's spoilers. However, a systemic or holistic analysis reveals numerous underlying causes: the impression given in the aircraft's manuals that mid-air deployment was not technically possible; the combination of two safety-critical functions in a single lever, thereby offering, to use Reason's phrase, an 'affordance' for error; the suspect strength of the DC-8's engine/pylon installation; the design of the wing tanks and wiring; the possibility that the DC-8's accelerated development programme (in comparison with that of the very similar 707) compromised the quality of the aircraft's design.

One could say, following Barlay's argument (1997: 119-120), that the design of the spoiler lever 'ensnared the pilot', or, using Tye's (cited in Hurst, 1976: 63-64) phraseology, that a '... failure of design ... allowed human error to occur too readily'. Certainly—to paraphrase Reason's memorable line—it would appear that the crew of the DC-8 were '... the inheritors of system defects created by poor design ... and bad management'. In pulling the lever while the DC-8 was airborne one might conclude that the First Officer '... added the final garnish to a lethal brew whose ingredients had been long in the cooking'. The misleading manuals were as much a part of the 'lethal brew' as was the design of the spoiler lever.

Of course, a more charitable interpretation of the design choices exercised by Douglas was that their designers had a right to expect such skilled and motivated workers as pilots (see, for example, Bressey (1976), Besco (1996) and Weir (1999), above) to be able to operate the aircraft's systems—even the dual-function spoiler lever—with an appropriate degree of concentration and dexterity. The Douglas engineers looked to the crew to achieve and maintain a suitable margin of safety with regard to the operation of the aircraft's various controls and systems *at all times and under all circumstances*. It could be argued, however, that in failing to factor in such variables as chronic fatigue or temporary aberration, the Douglas engineers had expected too much of the DC-8's crews.

The Trident disaster

The Trident disaster at Staines had a number of clearly identifiable immediate causes, such as the premature retraction of the aircraft's droops and the failure to maintain an adequate velocity throughout the take-off flight envelope. These were fundamental operator errors. But, as with the other disasters, there were a number of underlying causes. These may be listed under five rubrics: defects in design; problems of hierarchy; fitness to fly; adverse environmental conditions; and finally, procedural impositions on piloting.

Following Reason's discourse, it could be argued that the Trident's basic design contained three 'resident pathogens'. First the Trident's propensity, under certain conditions, to enter a deep or super stall from the basic stall condition. Secondly, the crew's ability to retract both the droops and flaps at the same time (resulting in a significant loss of lift). Thirdly a stall warning system with a reputation for 'false positives' (thereby reducing its credibility).

In the absence of a CVR one can only guess at the human interactions on the Trident's flight deck during the emergency. It might be assumed, however, that due to their relative inexperience neither the First Officer nor the 'monitoring pilot' would have been entirely at ease should they have considered it necessary to challenge Captain Key's decisions and/or actions. He was, after all, a 'veteran flyer'. It is also worth remembering that the emergency took place against the background of a recent acrimonious exchange in BEA's crew room (witnessed by the First Officer).

Assuming the post-mortem to be accurate, there is little doubt that at the time of the emergency Captain Key was experiencing serious health problems. These may have affected both his decisions and/or actions, the crew's relationship with their Captain and, possibly, the actions of the First Officer.

During much of the emergency the aircraft was in cloud. Consequently, there would have been few visual references to aid spatial orientation. Lastly, the Trident's crew were obliged to engage in a noise-abatement procedure during take off. This involved throttling back the engines. While not inherently dangerous, such a procedure may have had the effect of compounding operator error. It should be remembered that, throughout its take off, the Trident — possibly because of a misjudgement on the part of the crew — failed to attain the appropriate velocities. Throttling-back under these circumstances would have reduced safety margins. It might be argued that, had such an operational constraint not been placed on crews flying out of Heathrow, the Trident would not have stalled. To the extent that noise abatement flight profiles pacify local residents they may be deemed politically useful. But to the extent that noise abatement flight profiles (or, as Buck (1994: 3-4) would have it; 'hazardous noise reduction procedures') compromise safety, they may be adjudged antithetical to safe flying.

Thus it could be argued that the Trident disaster at Staines had a complex aetiology — an aetiology that, with respect to the design of the aircraft and its controls, could be traced back to the drawing boards of de-Havilland's engineers. Casting the causative net a little wider (following Reason's 'causal chains' discourse) it might be concluded that *ultimate* responsibility lay with those who granted the Trident design a Certificate of Airworthiness. Faith (1996a: 1) — uncharitably, perhaps — describes such actors as '... the shadowy figures ... who control the development of new aircraft, flight patterns and the discipline imposed on pilots'.

The MALEV reforms

The MALEV case is interesting for the way in which the airline's profound and chronic safety failings were habitually blamed on its aircrew. It is quite possible that a proportion of accidents and disasters were due to 'operator error'. But, as Barlay points out, there were numerous potential contributory causes. These included a largely incompetent senior management, poorly trained front-line engineering staff, a failure to implement management and maintenance systems appropriate to modern jet-powered aircraft and the expectation that ex-military fliers could readily adapt to commercial aviation. Consequently it can be suggested that the airline's safety problems originated in a melange of factors, some obvious, some not. It may be concluded both from this and the other case studies that many factors, from the most fundamental design decisions to lapses in the concentration of pilots or air traffic controllers, impact the safety of air transport. Or as Buck (1994: 3-4) puts it; 'This flying business is a big, complicated arena with many humans involved ...'.

Conclusion

This monograph set out to explore the connection between human performance and factors like operational requirements, the quality of relationships with work colleagues, design decisions and training regimes. Through systems theory it has been shown that operator errors may be facilitated or induced by factors — like poor design or operational procedures that, under certain conditions, compromise safety — that are beyond the control of pilots. Reason calls such factors 'latent errors'. Shaw calls them 'traps'. The case studies have demonstrated an important (and enduring?) characteristic of the air transport industry, namely the propensity of the public and some regulatory agencies to focus only on the *immediate* cause(s) of disaster, which most often involve some degree of pilot error. However, they have also demonstrated the utility of holism with regard to discovering the underlying cause(s) of disaster. The case studies would appear to confirm Reason's thinking on the aetiology of disaster in complex systems, namely that active operator errors may have their origins in 'resident pathogens' or 'latent errors'. The DC-8's First Officer could not have brought down his aircraft had the designers at Douglas incorporated a gate in the spoiler control that prevented airborne deployment. That is, this particular route to disaster could have been closed off if, through the exercise of foresight, the designers had made the mechanism *error resistant*. This, of course, reflects Norman's thinking on the matter of human error. Human error, he asserts, has its origins not in 'clumsiness' but in 'poor conception and design'. Norman's argument is that good design may reduce the chance of human error. However, it could be argued — on the basis of such incidents as the 1994 Aeroflot Airbus disaster — that even good design cannot protect against wilful dereliction of duty.

Serendipity is another factor beyond the control of the designer or regulator: On July 4th 2000 a fully serviceable MALEV Tu-154 carrying over 90 people came in to land at Thessaloniki Airport with its undercarriage up. ATC radioed the pilots, who attempted to climb away. The manoeuvre was only partially successful. The aircraft touched the runway and was damaged. After a go-around the crew landed the aircraft safely. At first sight this would appear to be an obvious case of pilot error. Certainly the crew's failure to lower the undercarriage was the immediate cause of the accident. But, as Doyle (2000: 9) asserts, there may have been contributory factors. First, the closure of the airport's usual runway, which required a 'late turn' onto a second (military) runway. Secondly the fact that another passenger jet '... cleared the runway just before the Tu-154 arrived at the threshold'. These factors '... may have distracted the crew'. Neither factor fell within the crew's orbit of influence or control.

The theme of good design (of hardware, software, training and operational procedures) as a bulwark against operator error (and potential disaster) is at the core of the relatively new discipline of human factors engineering or human-centred design. In the United States this approach to error mitigation and/or prevention has received the backing of the FAA, NASA and the DoD who, through their *National Plan for Civil Aviation Human Factors*, have sought to promote human factors engineering throughout that country's National Airspace System (NAS). The Plan, which focuses on such activities as '... automation ... selection and training ... human performance assessment ... information management and display [and] bioaeronautics' (1995: 3), is justified in the following terms:

> Many human errors that lead to accidents can be attributed to unforeseen effects of new technologies, new operational procedures and changing organisations. Aviation exists in an extremely complex and dynamic environment, and many factors can affect safety. As these factors change, the consequences of these changes on pilots, cabin crews, air traffic controllers, maintenance personnel, and airline operations personnel must be understood and dealt with before they cause accidents (1995: 32).

gued that this analysis reflects most of the factors highlighted by the
The Comet accident could, in part, be attributed to 'an unforeseen
w technology': the low-powered centrifugal turbojets required that

pilots fly the aircraft within a narrow flight envelope during take-off. Even if this was achieved (under all circumstances) the power plants struggled to get the Comet airborne (despite the aircraft's weight-saving design). It could be said that the Trident disaster—in part, at least—had its origins in the power-reductions required by Heathrow's noise-abatement regime. The MALEV crashes during the 1970s were, according to Matyasy, partly attributable to the airline's failure to adapt its organisation and practices to a new generation of passenger aircraft. To the extent that the *National Plan*'s approach to the issue of air safety reflects a systemic analysis of the aetiology of air disasters (perhaps including some of those described above) it could be said that its sponsors are engaged in isomorphic learning. They are trying to develop a long-term safety strategy on the basis of past successes *and failures*. As Toft and Reynolds (1997: 115) might say, they are seeking to exercise 'active foresight'.

For its part, the FAA has begun to confront some unpalatable truths. For example, that the much-vaunted cockpit automation of the 1990s (that saw the introduction of computerised flight management systems (FMSs)) may, unless designed and introduced with close reference to human capabilities and preferences, act to reduce safety. According to Learmount (2000a: 4) the FAA's research '... revealed that greater automation ... could produce crew confusion' and that '... the FMS created new types of pilot error'. According to O'Leary (2000) modern 'glass cockpits' (flight decks where tasks are highly automated and data is displayed on multi-function screens) have a detrimental effect on situational awareness by reducing vigilance and cross checking. Here we have further evidence that pilot 'error' may be induced—perhaps one could even say prescribed—by choices exercised by persons removed in time and space from the operational environment. Human error—by design. Quod erat demonstrandum?

References

Aviation Safety Network (1996) http://aviation-safety.net/database/1996/960206-0.htm

Baldry, D. (1980) *Piston Airliners Since 1940*. London: Phoebus.

Barlay, S. (1997) *Cleared For Take-Off—Behind the Scenes of Air Travel*. London: Kyle Cathie.

Beaty, D. (1991) *The Naked Pilot—The Human Factor in Aircraft Accidents*. London: Methuen.

Bordoni, A. (1997) *Airlife's Register of Aircraft Accidents*. Shrewsbury: Airlife Publishing Ltd.

Brookes, A. (1996) *Flights to Disaster*. Shepperton: Ian Allan Publishing.

Buck, R.N. (1994) *The Pilot's Burden—Flight Safety and the Roots of Pilot Error*. Ames: Iowa State University Press.

Campbell, R.D. and Bagshaw, M. (1999) *Human Performance and Limitations in Aviation*. Oxford: Blackwell Science.

Chant, C. and Batchelor, J. (1979) *Jet Airliners—An illustrated history of jet travel*. London: Phoebus.

Cullen, S.A. (1998) Aviation suicide: A review of general aviation accidents in the UK, 1970-96. *Aviation, Space and Environmental Medicine*, 69 (7).

DD Video (1991) *Airliner: Boeing 747*. North Harrow: DD Video

Doganis, R. (1991) *Flying Off Course—The Economics of International Airlines*. London: Routledge.

Doyle, A. (2000) 'Malev Tu-154 touches down in gear-up go-around'. *Flight International*. 15-21 August: 9

Edwards, M. (1974) *The Aviator's World*. London: David and Charles.

Evans, J. (1997) *Is It On Autopilot?* Shrewsbury: Airlife Publishing Ltd.

FAA, DoD, NASA (1995) *National Plan for Civil Aviation Human Factors: An Initiative for Research and Application—First Edition*. Washington DC: FAA.

Faith, N. (1996a) *Black Box: Why Air Safety is No Accident*. London: Boxtree.

Faith, N. (1996b) *Black Box: The final investigations*. London: Boxtree.

Flight International (1999) Designer Error. *Flight International*, 11-17 August.

Flight International (2000a) Promising Future. *Flight International*, 25-31 January.

Flight International (2000b) Airbus criticised over cowl incidents. *Flight International*, 18-24 July.

Francillon, R.J. (1979) *McDonnell Douglas Aircraft Since 1920*. London: Putnam.

Gero, D. (2000) *Aviation Disasters*. Sparkford: Patrick Stephens Limited.

Gilbert, J. (1970) *The Great Planes*. London: Ridge Press.

Godson, J. (1970) *Unsafe at Any Height*. London: Anthony Blond.

Greatorex, G.L. and Buck, B.C. (1995) Human Factors and Systems Design. *GEC Review*, 10 (3).

Hensser, H. (1953) *Comet Highway*. London: John Murray.

Hood, C. and Jones, D.K.C. (eds.) (1996) *Accident and Design—contemporary debates in risk management*. London: UCL Press.

Howlett, P. (1994) The 'Golden Age'. In Johnson, P. (ed.) *20th Century Britain—Economic, Social and Cultural Change*. London: Longman.

Huenecke, K. (1997) *Jet Engines—Fundamentals of Theory, Design and Operation*. Shrewsbury: Airlife Publishing Ltd.

Hurst, R. (ed.) (1976) *Pilot Error—A professional study of contributory factors*. London: Crosby Lockwood Staples.

Jackson, A.J. (1978) *De Havilland Aircraft Since 1909*. London: Putnam.

Job, M. (1994) *Air Disaster—Volume 1*. Fyshwick: Aerospace Publications.

Jones, L. (2000) CAA slated over aircraft safety. *The Daily Telegraph—Telegraph Travel*, April 1: 2.

Jones, M.A. (1983) *The Limits of Liberty—American History 1607-1980*. Oxford: Oxford University Press.

Kermode, A.C. (1972) *Mechanics of Flight*. London: Pitman.

Kingsley-Jones, M. (2000) Ageing Airliner Census 2000. *Flight International*. 13-19 June.

Kirkus Reviews (1993) reproduced at (2000) http://www.amazon.com/exec/obidos/ts/book-reviews/0201626.../102-1065115-449771. June 30.

Last, S.R. (1989) Who Is In Control? *The Log*. Harlington: BALPA. February.

Learmount, D. (1999) Safety Surprises. *Flight International*. 28 July - 3 August: 31-35.

Learmount, D. (2000a) The Year (and the 1990s) in Review. *Proceedings of the 12th annual European Aviation Safety Seminar, EASS, Safety: Beginning at the Top, March 6-8, 2000, Amsterdam, Netherlands*. Alexandria: Flight Safety Foundation.

Learmount, D. (2000b) Safety inspections reveal extent of danger airlines. *Flight International*. 6-12 June.

MacPherson, M. (1998) *The Black Box—Cockpit Voice Recorder Accounts of In-Flight Accidents*. London: Harper Collins.

Miller, R. (1984) Accident Analysis The Modern Way. *The Log*. Harlington: BALPA. August.

Norman, D.A. (1988) *The Psychology of Everyday Things*. New York: Basic Books.

Norman, D.A. (1993) *Things That Make Us Smart: Defending Human Attributes in the Age of the Machine*. Reading, Mass: Addison-Wesley Publishing Company.

Norman, D.A. (1997) People propose, science studies, technology conforms. Reproduced at (2000) http://www.amazon.com/exec/obidos/ts/book-reviews/0201626.../102-1065115-449771. June 30.

O'Leary, M. (2000) Situation Awareness and Automation. *Proceedings of the HCI Aero 2000 Conference*. Toulouse, France. September.

Perrow, C. (1999) *Normal Accidents*. New Jersey: Princeton University Press.

Reason, J. (1990) *Human Error*. Cambridge: Cambridge University Press.

Sagan, S.D. (1993) *The Limits of Safety — Organisations, Accidents, and Nuclear Weapons*. New Jersey: Princeton University Press.

Shkrum, M.J., Hurlbut, D.J. and Young, J.G. (1996) Fatal light aircraft accidents in Ontario: A five year study. *Journal of Forensic Sciences*. 41 (2).

Snook, S.A. (2000) *Friendly Fire: the accidental shootdown of U.S. Black Hawks over Northern Iraq*. New Jersey: Princeton University Press.

Tarry, C. (2000) Time to Break the Cycle. *Airline Business*. June.

Taylor, J.W.R. (1965) *Airliners of the World*. Britain: Odhams Books Ltd.

Taylor, M.J.H. (1997) *The World's Commercial Airlines*. London: Regency House.

Tegler, J. (1999) B-58 Hustler — Convair's Mach 2 Bomber. *Flight Journal*. 4 (6) December.

Toft, B. (1992) The Failure of Hindsight. *Disaster Prevention and Management: An International Journal*. 1 (3).

Toft, B. and Reynolds, S. (1997) *Learning from Disasters: A Management Approach*. Leicester: Perpetuity Press.

Vaughan, D. (1996) *The Challenger Launch Decision*. Chicago: University of Chicago Press.

Weir, A. (1999) *The Tombstone Imperative — The Truth About Air Safety*. London: Simon and Schuster.

Williams, T.I. (1987) *The Triumph of Invention — A History of Man's Technological Genius*. London: Macdonald Orbis.

Wynn Jones, M. (1976) *Deadline Disaster*. London: David and Charles.